Boeing 737

The World's Most Controversial Commercial Jetliner

Graham M Simons

AIR WORLD

AIR WORLD

Boeing 737

The World's Most Controversial
Commercial Jetliner

First published in Great Britain in 2020 by Air World Books,
an imprint of Pen & Sword Books Ltd,
Yorkshire - Philadelphia

ISBN: 9781526787231

Typeset in 12pt Times by GMS Enterprises
Printed and bound in India by Replika Press Pvt Ltd

Pen & Sword Books Ltd incorporates the imprints of Air World Books, Pen & Sword Archaeology, Atlas, Aviation, Battleground, Discovery, Family History, History, Maritime, Military, Naval, Politics, Social History, Transport, True Crime, Claymore Press, Frontline Books, Praetorian Press, Seaforth Publishing and White Owl.

For a complete list of Pen & Sword titles please contact:

PEN & SWORD BOOKS LTD
47 Church Street, Barnsley, South Yorkshire, S70 2AS, UK.
E-mail: enquiries@pen-and-sword.co.uk
Website: www.pen-and-sword.co.uk

Or

PEN AND SWORD BOOKS,
1950 Lawrence Roadd, Havertown, PA 19083, USA
E-mail: Uspen-and-sword@casematepublishers.com
Website: www.penandswordbooks.com

Contents

Acknowledgements

A book of this nature would not have been possible without the help of many people and organisations too numerous to individually mention. That said, back in 1974 the world became aware of 'Deep Throat', the secret informant who provided information in 1972 to Bob Woodward, who shared it with Carl Bernstein. Woodward and Bernstein were reporters for *The Washington Post*, and Deep Throat provided key details about the involvement of US President Richard Nixon's administration in what came to be known as the Watergate scandal. Jump forward nearly fifty years, and for this book very special thanks should go to my own secret informant who goes under the pseudonym of 'BJ' for providing a great amount of information from within Chicago and Seattle.

I am indebted to many people and organisations for providing photographs for this story, but in some cases it has not been possible to identify the original photographer and so credits are given in the appropriate places to the immediate supplier. If any of the pictures have not been correctly credited, please accept my apologies.

I would however, like to offer my special thanks the Fellows, Members and Associates of the Worldwide Society of Civil Aviation Enthusiasts (WSCAE) - many of whom work in the global aviation industry as flight or ground crew or indeed 'on the spanners' - for letting me have free access to their collections, including Dee Diddley, Ameen Yasheed, Wan Hunglo, Tsai Ing-Wen, John Hamlin, Bob Lehat, Kirk Smeeton, Ian Frimston, Carrie Åberg, Simon Peters, Robin Banks, Rick O'Shea, Hugh Jampton, Peter Clegg, Matt Black, David Lee, Emma Dayle, Dr Harry Friedman, Trymore Simango and Phil McCrackin.

Introduction

The crisis surrounding the crash of two 737 MAX jetliners was the biggest challenge Boeing management faced in many years. However, it was not the company's darkest hour; in fact, it was not even the first time that 737s crashed due to a design flaw.

Boeing has faced challenges to its survival in the past. The human cost of losing two 737s to avoidable accidents should not and cannot be minimised - hundreds of lives were tragically cut short, and families torn apart. However, the evidence from past crises suggests Boeing will survive and thrive again, probably sooner than many are expecting - thanks in part to the hard lessons the company has learned from its latest challenge.

The company was founded by William E 'Bill' Boeing and Commander G Conrad Westervelt in 1916, (see *Boeing 707 Group*, Pen & Sword 2017 by the same author) just in time to benefit from a surge of military orders during World War One. After the war, Boeing won lucrative airmail contracts from the federal government, and assembled a sprawling aviation conglomerate - its holdings included engine-maker Pratt & Whitney and what became United Airlines - aimed at dominating US aviation.

In 1934 the Roosevelt Administration cancelled the company's airmail contracts and broke up what it saw as an emerging monopoly. Bill Boeing was so upset by these moves that he quit the company that bore his name and sold all his stock. Unlike competing companies begun by aerospace pioneers such as Donald Douglas, Glenn Martin and James McDonnell, Boeing lost the driving force of its founder early on.

That could have been fatal for the enterprise - many aircraft companies failed during depression years - but it taught Boeing to operate without a visionary at its helm. Most of the companies whose founders stuck around later proved incapable of competing when that visionary faltered or died.

While Washington was prosecuting its trust-busting campaign against Boeing, the company introduced one of the most revolutionary products in its history. The Boeing 247 today is considered the first modern airliner. It was a monoplane rather than a biplane. It was all-metal rather than containing wood or canvas in its design. It had a streamlined shape and retractable landing gear to minimise drag. It even had a wing de-icing system.

The 247 was an impressive technological achievement, but few were sold. The airliner was eclipsed by the even more advanced Douglas DC-3, which could carry a larger number of passengers. In the end only seventy-five Boeing 247s were built, sixty which went to the company's airline unit that Washington would soon force it to divest.

Boeing and its competitors were rescued from the Great Depression by World War Two. Even before America entered the war, Washington launched a vast buildup of US airpower that multiplied the company's revenues. Boeing built the B-17 and B-29 bombers. But within weeks after victory was won in 1945, the government started wholesale cancellation of military aircraft contracts, resulting in around 70,000 Boeing workers losing their jobs. Boeing and its rivals thus entered the postwar period severely diminished and with uncertain business prospects.

Boeing transformed the B-29 airframe into a commercial airliner after the war, but what really lifted its prospects was the coming of the Cold War and the advent of jet engines. The Cold War generated demand for new Boeing bombers, most notably the jet-powered B-47 and B-52. Boeing also built hundreds of jet-powered tankers to support the bomber fleet, and work on that programme contributed to the development of its 707 jetliner via the Model 367-80 - a late-comer to civil aviation after the British, French, Canadians and even the Russians.

By the late 1960s, Boeing was the global leader in jetliners, having developed the single-aisle 727 similar to the earlier British DH.121 Trident and then the 737 - itself a concept from the British of fifteen years or so earlier - that was to become the most widely used commercial transport in history, and the 747 that held the world record for passenger-carrying capacity over four decades. Both designs were eventually eclipsed by Airbus products with their A320 and A380 respectively - a simple fact that grated on American egos.

Development of the 747 left the company heavily in debt. An economic recession caused orders for commercial aircraft to dry up just as military demand generated by the Vietnam War was softening. Between 1968 and 1971 the number of workers at Boeing's commercial aircraft unit fell over seventy-five per cent. Company finances were stretched to the breaking point, and bankruptcy was barely averted.

In 1991 a 737 crashed on approach to Colorado Springs after inexplicably rolling to the right and then going into a steep dive. Everybody on board died. In 1994 it happened again near Pittsburgh, with the plane this time rolling to the left before pitching down. Once again, everybody on board was killed. These 737 accidents did not provoke the same crisis atmosphere as 737 MAX mishaps did, because they were much more widely separated in time and thus did not create a media frenzy of reporting. However, a four-year investigation by the National Transportation Safety Board after the 1994 crash concluded that a malfunction in the rudder had caused it to reverse direction, causing both tragedies. The US Federal Aviation Administration ordered modification of a key mechanism and mandated new training procedures so pilots could cope with unanticipated shifts in the flight control system. Globally, the entire fleet was discreetly modified.

The growing commercial threat of the European aerospace company commonly called Airbus to Boeing cannot be understated. The inroads that the Europeans were making into Boeing's global market and especially their own national market was dramatic. Something had to be done and done quickly. A product was needed that was more attractive than its rivals were offering. This was the 737 MAX.

Writing this book has proved to be challenging - reading it may be the same. The difficulties, as usual with much of my work, springs from the differences of the so-called 'common language' British and Americans share. Color becomes colour, program becomes programme, and of course, American phrasing is often different from English. Then there is the dreaded use of plane instead of aircraft; I don't care what anyone says, a plane is a cutting tool used to smooth wood in my books!

This brings me to the conundrum of whether to use Imperial or metric units of measurement? Invariably there are times when I must use both, but by and large my writing rules are simple: the aircraft was designed using Imperial units, so I use Imperial measurements. I am English, so I write in that language; however, as a sign of respect to that nation, if I am quoting an American, I use their spelling and phrasing in any quotation.

Another difficulty is that for much of this book two independent storylines are in place - the chronology of the 737, which in itself is somewhat convoluted, and the commercial and political events that were swirling around it. Inevitably this has produced a disjointed main storyline, which I have tried to at least partially resolve by telling the story in a series of almost stand-alone chapters. Unfortunately it does mean that some photographs do not mesh into place with the main body text.

Boeing and its corporate antecedents have faced many other challenges, such as the end of the Cold War which severely reduced demand for military aircraft, and the rise of a European jetliner producer that eventually claimed half of the global market. The enterprise as it exists today was shaped by the stresses these recent events created.

Can Boeing survive the 737? In order to try and answer that question, it is necessary to go back and look at all that went on, where the airliner came from and the events that have happened.

Graham M Simons
Peterborough.
September 2020

Chapter One

Project Origins

In late 1958 the Boeing Commerical Aircraft Company announced a design study for '...a twin-engined feeder airliner to complete the family of Boeing passenger jets' to supplement their trijet 727 on short and thin routes.

Boeing wanted a true short-haul jet to compete with the Caravelle, BAC One-Eleven and DC-9 but was way behind them. The DC-9 was about to fly, the One-Eleven was well into its flight test programme and the Caravelle had been in service for five years. Also in Europe, Fokker Aircraft were about to produce their F.28. Clearly, Boeing had some catching up to do.

Boeing - and indeed the USA - had been late to enter the jet-airliner market, being beaten by just about all the other manufacturers globally.

With the introduction of the Boeing 707, derived from the KC-135 air refuelling tanker for the USAF, itself derived from the Boeing Model 367-80, demand for jets on shorter routes was soon a reality. Indeed, propeller-driven airliners were fast becoming an anachronism in the minds of the traveling public - long before the economics of jets justified them on short routes.

In May 1958, before the first 707 was delivered to Pan American, John E. 'Jack' Steiner, was appointed to head a planning group for a Boeing short-range jet. At that time, Boeing engineer Maynard Pennell, who had cut his teeth on the B-29, and headed the 707 programme as senior project engineer, was in charge of preliminary design. The two men had great respect for each other's ability, sharing a common vision - the maturation of the jet airliner business.

It was a fortunate coincidence that Steiner and Pennell had come together at

Key figures in the Boeing 727 and 737 projects.

Left: Joseph Frederick 'Joe' Sutter (*b*. 21 March 1921, *d*. 30 August 2016)

Right: John Edward 'Jack' Steiner (*b*. 1917, *d*. 29 July 2003)

British Aircraft Corporation BAC 1-11 G-AVOF, a series 416EK, seen in the colours of Cambrian Airways but with Gulf Air titles. *(author)*

a propitious moment in history. It was Pennell, perhaps more than others, who led the thrust into the commercial jet age.

When Steiner was given the short-range jet assignment in 1958, the concept had a number - the 727 - but not much more. Literally hundreds of design variations had been tried, and it was beginning to look like an endless journey. To keep the airliner as small as possible, they had concentrated on two-engined versions, but two engines under the wings posed a serious load-and-balance situation, and two on the back added excessive structural weight to the fuselage.

Discussions with the airlines did not provide much to go on. Uncertainty as to what the next step should be was predominant. United Airlines, regarded as a bellwether of the industry in the USA, used Denver as a hub on its transcontinental flights, and a one-stop airplane would have to be good for mile-high Denver. Performance requirements for a twin, with one engine out, would be very difficult to meet.

Boeing found TWA receptive to a twin - but only lukewarm. In a phone conversation with Bob Rummel, TWA engineering Vice president, Steiner listened intently, when Rummel, half-joking, said, ''Why don't you compromise on three engines and make a good airplane?'

Boeing's interest in a three-engine airliner stood at zero. There were more problems than with the twin. Where to put the third engine - if indeed the right-sized engines were available - or even on the drawing boards?

The design studies for the 727 continued, and after still another review, engineering could not find anything that made sense and was beginning to harden on the four-engine concept, a machine slightly smaller than the 720. The sales department baulked, insisting on a truly small airliner. Bob Rummel of TWA was still pushing for three engines.

By 1959, the marketing department presented a depressing outlook. 'The market is not yet ripe for exploitation,' their report read. They added that whether the airplane was built by Boeing or Douglas, it would probably have to be priced well above the figures used in the study, and close to the price of the 720. 'In short, it appears that a small jet is not economically possible.'

With new urgency, Steiner and Pennell got Bill Allen's approval for a two-month, all-out evaluation programme, culminating in a go or no-go recommendation to the corporate offices.

In late June, the puzzle was complicated further when Eastern indicated their preference for three engines. American was adamant on two, and Douglas publicly announced its decision to build a four-engined DC-9.

Taking stock, Steiner concluded that Douglas was ahead with their four-engined version, with a tie-up with Pratt and Whitney for engines. In England, de Havilland was ahead on a three-engined design, and a three-engined airliner seemed to be common ground that might get all the major airlines together. But there were no engines of the right size.

Rolls Royce was building the engines for the three-engined de Havilland Trident. Steiner needed more power, urging Rolls to design a larger one. Desiring to get into the US market, Rolls agreed.

The team considered a three-engined airliner with two mounted on the wings and one in the tail. A breakthrough on the lift problem had been achieved, and when the test results were all digested, the aft-mounted engine design looked 'surprisingly good'.

Just when it appeared that the technical problems were yielding to a solution, the tooling estimate of 1.5 million staff-hours shot up to 5 million. A 727 production decision was still on hold.

When British European Airways signed a contract for twenty-four three-engined Tridents at the end of August 1959, with all mounted aft, and a 'T' tail, Pennell decided that the Trident was the competition to beat. Boeing looked at the specifications for the Trident and simply put all their efforts into designing a better machine. Location of the horizontal tail was the crucial parameter.

Engines mounted on the tail called for still another extensive wind tunnel testing programme. By May 1960, Joe Sutter had the results. The low tail came out negative and the 'T' tail was in. Finally, the 727 had a configuration. The tail design, with sweepback, provided an unforeseen benefit, having the effect of lengthening the machine, which increased the latitude for load-and-balance.

Concurrently, the horizontal stabiliser was smaller, trading off for the more substantial structure required for the aft-mounted engines.

It appeared that the time was ripe for the Boeing engineers to once again mine the rich vein of British invention and

Above: Robert W Rummel (*b.* 4 August 1915, *d.* 17 October 2009) of TWA

Left: Maynard Lyman Pennell (*b.* 12 April 1910, *d.* 22 November 1994)

Right: William McPherson 'Bill' Allen (*b.* 1 September 1900, *d.* 28 October 1985) President of Boeing from 1945 to 1968.

'Continental Golden Jet' DH.121 Trident G-AVYB of Channel Airways, seen at London Gatwick. The Tridents - apart from those sold to British European Airways, also sold in small numbers to other airlines. *(author's collection)*

innovation, having already once drank from that particular well with the freely provided data from de Havilland's on podded engines and metal fatigue following the disasters that struck the world's first jet airliner, the Comet. Boeing found themselves in a position of unique advantage. With the British tri-jet already committed to the factory, Boeing found itself sitting in the catbird seat with a 'paper airplane'; it was no great task to design a superior product. For those that do not speak American, 'The catbird seat' is an American English idiomatic phrase used to describe an enviable position, often in terms of having the upper hand or greater advantage in any dealing among parties.

That is exactly what happened. Engineering design for production - with no prototype - was committed on 10 June 1960. The first 727-100 was delivered to Eastern Airlines on 23 October 1963.

Joe Sutter agreed to go to Bill Allen with a proposal. 'I think it would be worth putting up a half-million to take a ninety-day look at it,' he told the President of the company. 'I don't believe there is one chance in ten that we can come up with anything that makes sense, but I think we ought to do this to make sure. The DC-9 doesn't have leading-edge flaps; we may be able to pick up some other advantages by reason of our later start.' Allen agreed, and the Model 737 programme study began on 8 May 1964.

Intensive market research and design studies yielded plans for a fifty-to-sixty-passenger airliner for routes between fifty and one thousand miles long and to be able to break even at a thirty-five per cent load factor. The initial design featured a pair of podded engines on the aft fuselage and a T-tail in a similar manner to the 727, but with five-abreast seating similar to what the competition had already decided upon.

It did not take Boeing long to realise that if they chose to go to six-abreast, it would provide significant commonality with the 707 and 727 fuselages. Having the same cross-section also provided for standardisation of cargo containers.

However, in the airliner business, no advantage comes without a price. The seventeen-inch wider cross-section, with its increased drag, resulted in a fifteen-mile-per-hour penalty in cruising speed compared to the DC-9.

Location of the 737 engines was open to the option of wing or body mounting. Sutter wanted to take another look at the wing-mounted design. In theory, the larger engines should not produce the same interference effects with the airflow that forced the 707 to go to pods. Testing proved the concept to be sound, and 1,200 pounds of weight was saved by mounting on the wings.

After further consultation with potential customers, Boeing's Joe Sutter, who had previously worked on many commercial airliner projects, including the 367-80 'Dash 80', 707 and 727, decided that a better option would be to put the engines under the wings to lighten the structure, enabling fuselage widening for six-abreast seating.

Legend has it that Joe Sutter knew that Boeing was competing in broadly the same market as the Sud Aviation Caravelle, BAC One-Eleven and Douglas DC-9 and they needed to come up with something different. Sutter took some scissors and cut up a drawing of the initial design for the 737 which also had a T-tail with aft-mounted engines. He began moving the engines around to find a better layout. But putting the engines on struts under the wing like the 707 would block boarding access to the main cabin door on the shorter fuselage of the 737.

'I slid the cut-out tight under the wing and felt a sudden flash of excitement. Instead of mounting the engines away from the wing on struts, why not mount them hard against the underside of the wing itself?'

The wing-mounted engines gave the advantages of reduced interference drag, a better C of G position, quieter aft cabin, more usable cabin space at the rear, fore and aft side doors, reduced landing gear length and kept the engines low to the ground for easy ramp inspection and servicing; they also required less pipework for fuel and bleed air. The weight of the engines would also provide bending relief from the lift of the wings. However, this benefit was somewhat overestimated, resulting in a set of wings which failed in static tests at ninety-five per cent of maximum load, so the wing had to be redesigned.

Many thickness variations for the engine attachment strut were tested in the wind tunnel and the most desirable shape

DC-9-31VH-CZA of Ansett-ANA *(Ansett-ANA)*

SA de Transport Aérien (SATA) or SA de Transport Aerien Genève, was a Swiss airline founded in June 1966, with its head office at Geneva Airport, and with its base in Geneva. It finished its activities in 1978. It started as an air-taxi operator and evolved to passenger and cargo charters to points in Europe, the USA, South America and the Caribbean. Here one of its three Sud Aviation Caravelles, HB-ICQ is seen at Geneva. *(Matt Black collection)*

for high speed was found to be one which was relatively thick, filling the narrow channels formed between the wing and the top of the nacelle, particularly on the outboard side.

The disadvantage of wing-mounted engines was that the size of the fin had to be increased for engine-out operation over centreline thrust aircraft. Also, due to the reduced ground clearance, the engines had to be almost an integral part of the wing, which in turn necessitated a short chord. The engines extended both forward and aft of the wing to reduce aerodynamic interference (further improved by the longer tailpipe of the target thrust reversers in 1969) and the straight top line of the nacelle formed a 'stream tube' - also known as a streamline flow over the wing to further reduce drag.

Initial worries about the low mounted engines ingesting debris proved unfounded; this had been demonstrated anyway by the Boeing 720B, whose inboard engines were lower than the 737's and had been in service for four years without significant problems.

Originally, the span arrangement of the airfoil sections of the 737 wing was planned to be very similar to that of the 707 and 727, but somewhat thicker. A substantial improvement in drag at high Mach numbers was achieved by altering these sections near the nacelle. The engine chosen was the Pratt & Whitney JT8D-1 low-bypass ratio turbofan engine, delivering 14,500 pounds of thrust. With the wing-mounted engines, Boeing decided to mount the horizontal stabiliser on the fuselage rather than the T-tail style of the Boeing 727. The final wing was a work of art as the specification required both good short-field performance and economy at altitude.

Overall, the wing-mounted layout had a weight saving of 1500 pounds over the equivalent 'T-tail' design and had performance benefits. A further advantage of the wing-mounted engine design was its similarity to earlier Boeings such as the 707.

It was by no means the first time this particular design concept had been attempted. Yes, it is correct to say that other aircraft manufacturers proceeded down the T-tail, rear-engined development line, and on entering this specific market - as with the 707 family - Boeing was lagging far behind its competitors when the 737 was launched. Rival airliners such as the British Aircraft Corporation BAC 1-11, Douglas DC-9, and Fokker F28 were already into flight certification.

However, there had been one previous, little-known British design that looked almost exactly like the 737 design that eventually appeared; this was the second prototype Vickers Viscount and took to the skies at least seventeen years before the 737. Known as the Type 663 testbed, it had two Rolls-Royce Tay turbojet engines. Construction began at Foxwarren, Surrey under revised Specification E.4/49 and transferred to Wisley Airfield in the same county for final assembly fitted with two Rolls-Royce RTa1 Tay turbojets to undertake control trials in support of the Vickers Valiant test programme. The Tay engine was a development of the Nene and produced 6,250 pounds of static thrust.

Three views of the Vickers Type 663 testbed VX217, built to Specification E4/49, that first flew in 1950. The resemblance of the 737 of some fifteen or so years later is uncanny. *(author's collection)*

Fokker F.28-4000 G-JCWW in in one of the hangars at Norwich Airport in a hybrid Air Anglia/Air UK colour scheme. *(author)*

The Type 663 first flew from the 6,600-foot grass runway at Wisley, piloted by Gabe Robb 'Jock' Bryce in RAF markings as serial VX217 - although it had been allocated the civilian registration of G-AHRG - on 15 March 1950. After initial manufacturer's trials, it was delivered to the Ministry of Supply for high altitude trials and operated from Seighford Airfield, Staffordshire, before appearing in public at the 1950 Society of British Aircraft Constructors (SBAC) air show at Farnborough airfield, Hampshire, including flying demonstrations by Vickers test pilot Brian Trubshaw.

Towards the end of the year, it was leased to Boulton Paul Aircraft Ltd, when it operated from Defford airfield, Worcestershire on electrically powered flying control trials, then was leased to Louis Newmark Ltd for trials and finally leased to the Decca Navigator Co. In 1957 the twin-jet was returned to the Ministry of Supply at Seighford airfield, Staffordshire, where on 31 December 1958 it was damaged beyond repair by fire due to a hydraulic failure in a wheel bay.

Promotion and Construction

In Europe, Boeing's international sales manager Ken Luplow had been discussing a 737 with the German flag carrier Lufthansa. The airliner appeared to be ideally suited to serving the multiple major cities in Germany.

Lufthansa could trace its history to 1926 when Deutsche Luft Hansa A.G. (styled as Deutsche Lufthansa from 1933 onwards) was formed in Berlin. DLH, as it was known, was Germany's flag carrier until 1945 when all services were terminated following the defeat of Nazi Germany. To create a new national airline, a company called Aktiengesellschaft für Luftverkehrsbedarf (Luftag), was founded in Cologne on 6 January 1953, with many of its staff having worked for the pre-war Lufthansa. West Germany had not yet been granted sovereignty over its airspace, so it was not known when the new airline could become operational. Nevertheless, in 1953 Luftag placed orders for four Convair CV-340s and four Lockheed L-1049 Super Constellations and set up a maintenance base at Hamburg Airport. On 6 August 1954, Luftag acquired the name and logo of the liquidated Deutsche Lufthansa for DM 30,000, thus continuing the tradition of a German flag carrier of that name.

On 1 April 1955 Lufthansa won approval to start scheduled domestic flights, linking Hamburg, Düsseldorf, Frankfurt, Cologne, and Munich.

International flights began on 15 May 1955, to London, Paris, and Madrid, followed by Super Constellation flights to New York City from 1 June of that year, and across the South Atlantic from August 1956. In August 1958 fifteen Lufthansa 1049Gs and 1649s left Germany each week to Canada and the United States, three 1049Gs a week flew to South America, three flew to Tehran and one to Baghdad. In parallel, the airline also initiated a marketing campaign to sell itself and West Germany. The challenges involved encouraging travellers to consider visiting the country in the wake of World War Two, as well as offering services to other nations via the Frankfurt airport hub. More specifically, Lufthansa's efforts shaped and reflected the development of a modern form of consumerism and advertising through the sale of air travel. By 1963, the airline, initially limited in its public relations efforts, had become a major purveyor of West Germany's image abroad.

With Lufthansa moving favourably toward the 737, Allen was in a dilemma - worse than on the 727. There, he had orders for eighty machines in hand with no competitor in the field. On the 737, there was strong competition and only one potential customer. Allen launched an intensive sales campaign to line up Lufthansa, United and Eastern.

By January 1965, a crisis had developed. Neither Eastern nor United would make any commitments. Lufthansa insisted on a Boeing yes or no as to the go-ahead with the 737, for they, Lufthansa, were ready and willing to buy DC-9s.

In February, the Boeing board was forced to face the problem head-on, with no new movement on the part of the US prospects. Although the market looked bleak, several members argued for a go-ahead, if only because holding off posed a still bigger risk. Fortunately, airline traffic was booming - and so they decided to proceed.

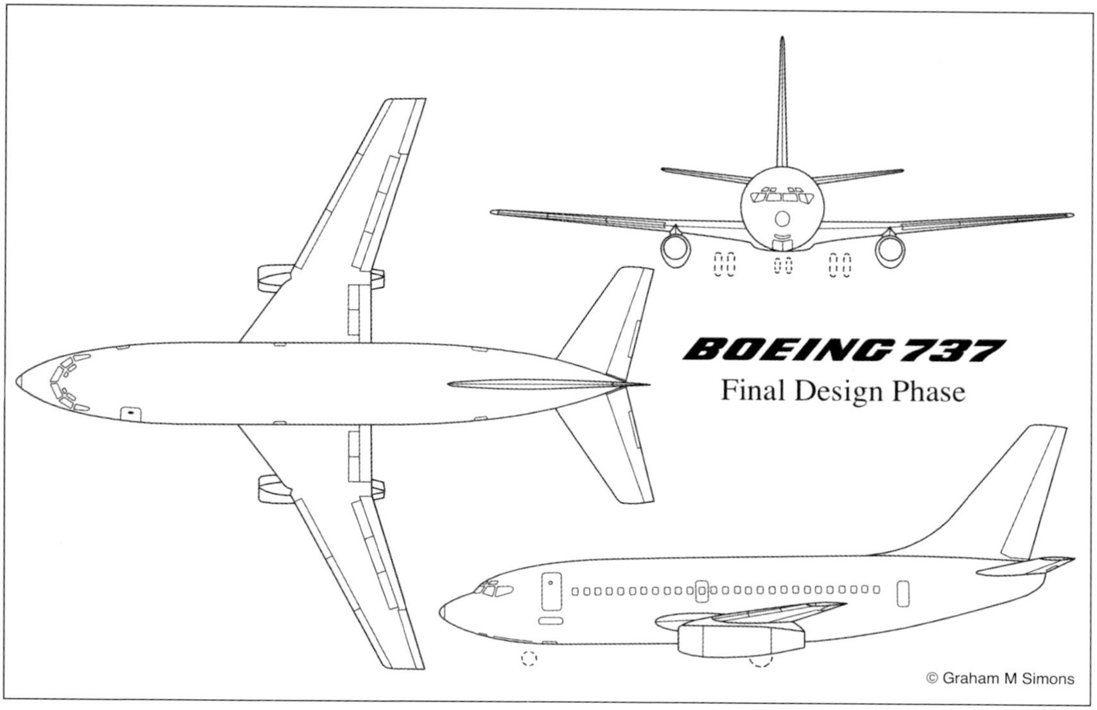

The 1965 Boeing Brochure for the 737-100 highlighted a number of features, some of which inadvertently played a major part in the ' problems' experienced by the type many years later.

'Maintainability

Maintainability successes achieved with the 727 are reflected in adoptions and refinements for the 737. Other improvements realized may make the 737 - from the standpoint of normal small airport operation - the most easily maintained airplane yet built.

'Eye Level' Maintenance

A unique characteristic of the 737 is its 'eye level' maintenance features. Nearly all maintenance requirements may be carried out at ground level without ladders, hoists, or other high lift devices. This means that the operator will not be burdened by the necessity of providing additional above-the-ground service equipment.

Engine Maintenance

The wing-mounted JT8D engines eliminate the need for high lift engine change equipment. Unlike other short range jet transports, the 737 engines may be changed at ground level with simple hand-hoists attached to the airplane structure. Side beams, designed for 727 engine changes, may be used with only minor modifications. Small airport equipment, presently in use, is all that's necessary for normal engine maintenance.

Systems Maintainability

Major systems are grouped at readily accessible locations for easy maintenance. Hydraulic units are located in the main wheel wellsreadily accessible for a mechanic standing on the ramp. Air-conditioning units under the wing are accessible through large doors.

Electronics equipment is stored in racks in a bay aft of the nose wheel well. Major control system components in the wing are accessible through leading-edge and trailing-edge openings. The APU in the tail end of the fuselage is cowled like an engine. The gas turbine APU and all its accessories are readily accessible for maintenance through cowl openings.

Departure Reliability

That low maintenance costs result in decreased operating expense should come as no surprise to the operator of Boeing airplanes. The 700 series jet transports have enjoyed historic success in this area. Each succeeding model from the 707-120 through the 727 has shown a marked increase in maintainability, and more significantly still, in despatch reliability. The 707/720 series achieved reliabilities of 97 per cent. The 727 has achieved a departure reliability of 98 per cent. The 737 is expected to show a departure reliability of 99 per cent.'

The brochure sang the praises of the delights passengers would find onboard.

'Passengers and crew board the 737 through a single forward entry door on the left side. A full-size aft entry door with or without the integral stair is available as an option. All entry and service doors are of plug-type design. Two emergency exit doors are centrally positioned in the passenger compartment with access to the wing. In combination with service and entry doors at each end, these emergency exits provide more than adequate escape egress.

'Airstairs carried directly under the forward entry door eliminate the need for airport stairways. The airstairs are stowed underneath the cabin floor, and a

An early cutaway showing the construction details of the 737. *(author's collection)*

plug-type door seals the airstair aperture.

'Six hundred and fifty cubic feet of cargo space is packaged in two compartments, one forward and one aft of the wing. Access to each of these areas is through large cargo doors opening on the right side.

'Immediately ahead of the forward cargo compartment is the electrical and electronic equipment bay. Cooling, as required, of shelf-mounted electronics packages is provided by a blower. Air-conditioning packs, accessible through large access doors, are centrally located forward of the main landing gear.

'The auxiliary power unit (APU) with built-in sound suppressors installed in the tail end of the fuselage supplies electrical energy and air for engine starting and air conditioning on the ground. In flight, the APU provides air-conditioning air or electrical power in the event of an engine shutdown. With electrical power, hydraulic power is also available from electrically driven pumps'.

Although the use of more modern materials have evolved over the years, the basic construction of the 737 remains the same. The fuselage is a semi-monocoque structure, meaning the loads are carried partly by the frames and stringers and partly by the skin. It is made from various aluminium alloys except for the use of

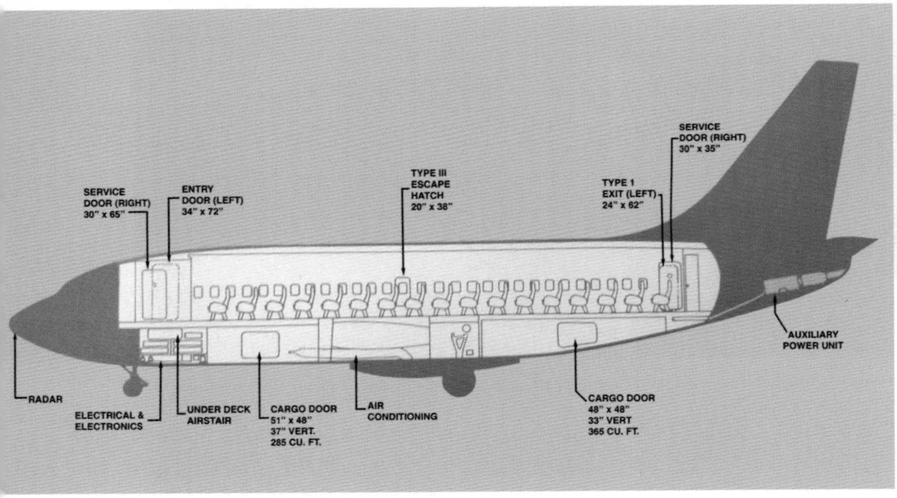

Above: the cover of the 1965 Boeing Sales Brochure.

Inset is from page 13 of the brochure, showing the interior layout.

Fibreglass for the radome, tailcone, centre and outboard flap track fairings. In later years, Kevlar was used for engine fan cowls, inboard track fairings behind the engine and nose gear doors. graphite/epoxy was used for the rudder, elevators, ailerons, spoilers, thrust reverser cowls, and the dorsal part of the vertical stabiliser.

Different types of aluminium alloys were used for different areas of the aircraft depending upon the characteristics required. The alloys are mainly of aluminium and zinc, magnesium or copper but also contain traces of silicon, iron, manganese, chromium, titanium and zirconium. The different alloys are mixed with different ingredients to give different properties. The fuselage skin, slats, flaps, stabilisers - areas primarily loaded in tension - were aluminium alloy 2024, a metal that was basically an alloy of aluminium and copper. This gave good fatigue performance, fracture toughness and slow crack propagation rate. The frames, stringers, keel and floor beams, wing ribs used aluminium alloy 7075, an alloy of aluminium and zinc, so as to provide high mechanical properties and improved stress corrosion cracking resistance.

On the 737-200, the bulkheads, window frames, landing gear beam made use of aluminium alloy 7079, an aluminium and zinc alloy, tempered to minimise residual heat treatment stresses. The wing upper skin, spars and beams made use of aluminium alloy 7178 a complex alloy of aluminium, zinc, magnesium and copper that gave a high compressive strength to weight ratio. The landing gear beam used aluminium alloy 7175, another complex alloy of aluminium, zinc, magnesium and copper that produced a very tough, very high tensile strength material. For the wing lower skin aluminium alloy 7055 was used. This was an aluminium, zinc, magnesium and copper alloy that gave superior stress corrosion.

The design life for the 737 was originally set at twenty years, 75,000 cycles or 51,000 hours; this was increased to 130,000 cycles in 1987.

To expedite development, Steiner made use of as much as possible of the 727 and 707 design in the 737, allegedly using sixty per cent of the structure and systems of the existing 727 in particular, the 'double-bubble' fuselage cross-section that was common to both designs. This gave not only cost savings in tooling commonality but also the payload advantage of six-abreast seating, one more than the DC- 9 or BAC-1-11 and allowed it to carry standard-sized cargo containers on the main deck. It gave the interior a spacious look and

Page 14 showed different interior layouts, that had some what today seem unusual features: single rod coat closets and an area for carry on baggage being good examples!

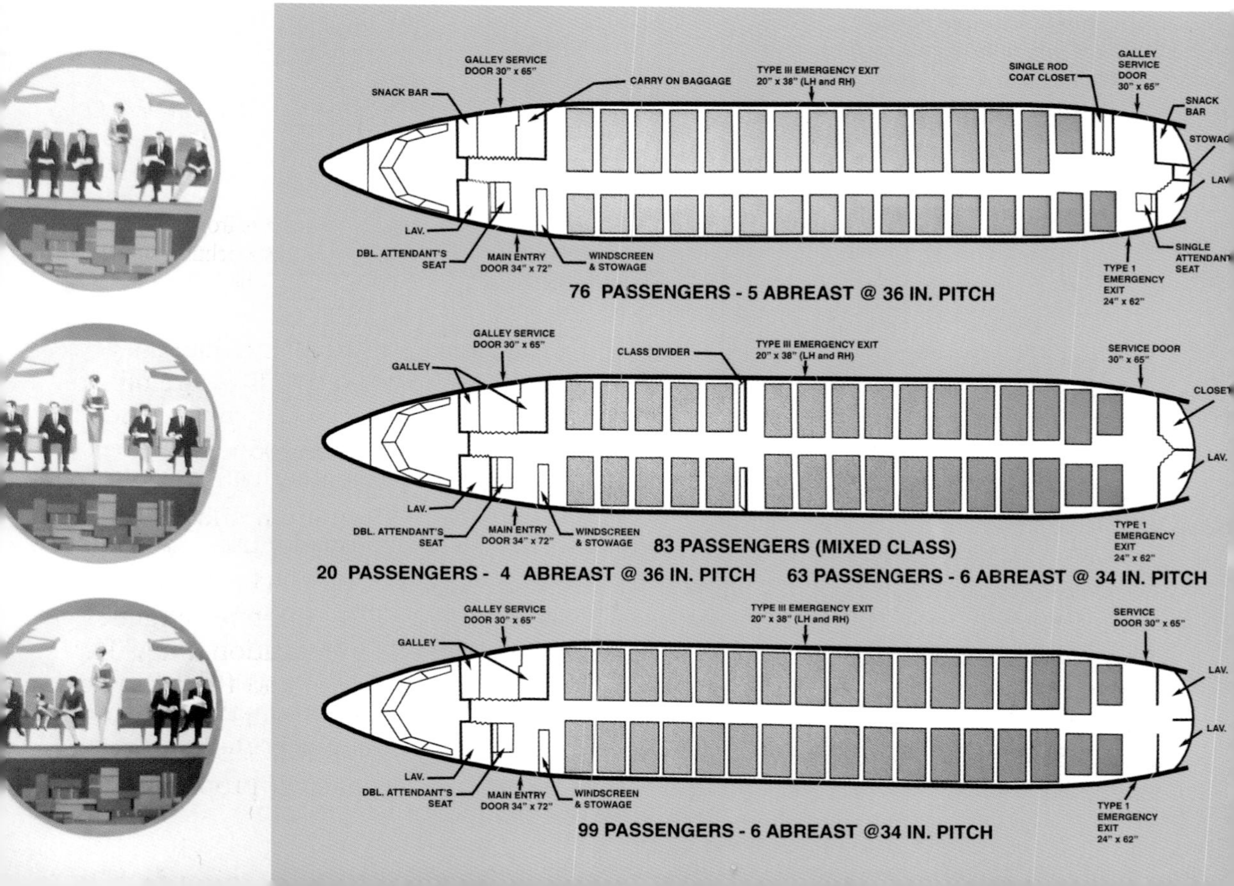

GROUND SERVICING

A principal reason for the 737's low operating cost is its unusually short servicing time.

The 737 has been designed for maximum freedom from ground support equipment and servicing units. The APU, started by the aircraft battery, supplies power for air conditioning (cooling and heating) on the ground, engine starting, and electrical power for lights, radio and other instruments. The 737 also incorporates its own passenger loading and unloading 'airstair,' thus eliminating the need for airport ground facilities.

Normally, the airstair is electrically powered. However, in the remote event that power failure occurs, the airstair may be operated manually.

The only service required for a minor thru-stop is baggage handling. Average minor stop ground time from 'airplane-parked' to 'ready-to-taxi' is estimated at thirteen minutes. No ground equipment except a baggage loader is required at minor thru-stops.

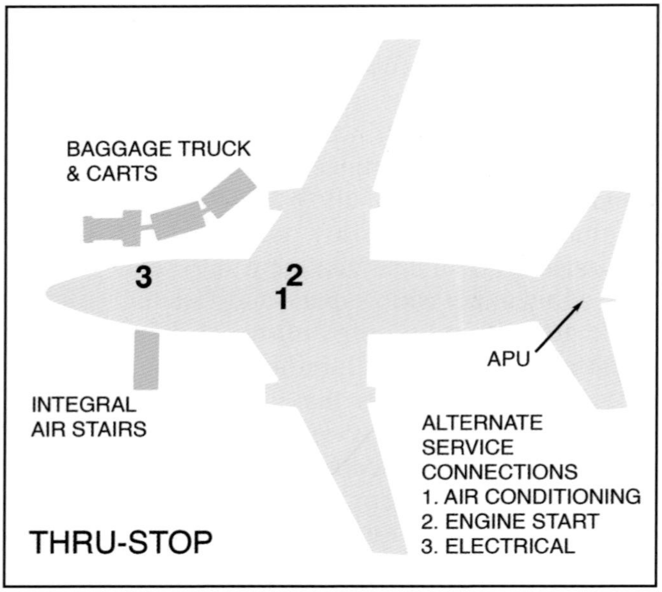

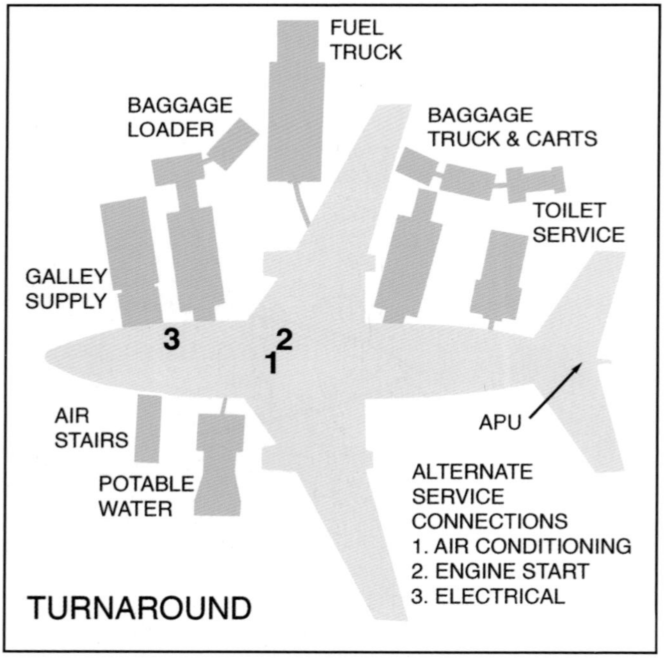

allowed Boeing to use standard cabin fittings, such as toilets and galleys.

The 737 wing was virtually a scaled down version of the 727 wing - the return to wing-mounted powerplants was dictated largely by the simple fact that in the smaller machine, a larger percentage of the total fuselage volume was needed for passenger seating. Aft-mounted engines ruled out a considerable volume that needed to be commericially available for the 737.

The 737 also made use of doors, leading edge devices, large parts of the nacelles, cockpit layout, avionics, components and other fittings. Using off-the-peg components was quick and cheap for both design and production and also helped pilots and engineers convert to the new type.

As the 737's fuselage design was derived from that of the Boeing 707, it inherited the distinctive 'eyebrow windows' used on the larger airliner, situated just above the windscreen.

The 737's main landing gear, under the wings at mid-cabin, rotates into wheel wells in the aircraft's belly. The legs are covered by partial doors, and 'brush-like' seals aerodynamically smooth (or 'fares in') the wheels in the wells. The sides of the tyres are exposed to the air in flight. 'hub caps' complete the aerodynamic profile of the wheels. It is forbidden to operate without the caps, because they are linked to the ground speed sensor that interfaces with the anti-skid brake system. The dark circles of the tyres are clearly visible when a 737 takes off, or is at low altitude.

737s are not equipped with fuel dump systems. The original aircraft were too small to require them, and adding a fuel dump system to the later, larger variants would have incurred a large weight penalty. Boeing instead demonstrated an 'equivalent level of safety'. Depending upon the nature of the emergency, 737s either circle to burn off fuel or land overweight. If the latter is the case, the aircraft is inspected by maintenance personnel for damage and then returned to service if none is found.

The 737 design was presented in

A shot of the rear fuselage of a 737 under construction that clearly shows the cross-section, a 'double-bubble' strengthened and braced by the cabin floor beams. *(Simon Peters Collection)*

October 1964 at the Air Transport Association maintenance and engineering conference by chief project engineer Jack Steiner, where its elaborate high-lift devices raised concerns about maintenance costs and dispatch reliability.

In Germany, at the Lufthansa Board meeting on 19 February 1965, chief executive Gerhard Hoeltje was reluctant - as the only airline customer - to recommend the 737. It would be easy for Boeing to drop the programme. With board members already arriving for the meeting, Hoeltje phoned to pin Boeing down. He wanted personal assurance from Bruce Connelly, vice president of the Transport Division.

Ken Luplow made an urgent call to Seattle. It was 10:00 am in Cologne, 1:00 am in Seattle. Rubbing the sleep from his eyes, Connelly gave the green light, and the Lufthansa board continued to deliberate. After breaking for a late lunch, they approved an order for twenty-one 737s. Within a week, Eastern announced their decision for the DC-9.

Success for the 737 was hanging by the United Airlines thread. Production with only Lufthansa as a customer appeared to be financial suicide.

The competitive battle centred around the perception of market growth and market share; and to whether the airliner with the greatest number of seats was the best revenue earner. Nobody wanted to fly empty seats.

With United involved in a major fleet planning programme to go all-jet by 1970, Boeing concentrated on a long-range programme of a 737/727 mix which would meet the airline's combined requirements. The advantage of already having the 727 in the United fleet was significant. To make up for the delay in getting the 737 into service, Boeing offered additional 727 machines at attractive terms, with the right to turn back the extra 727s when the 737s became available. The strategy worked.

In April, United announced a gargantuan order: forty 737s, twenty 727 passenger and six 727QC airliners. They leased an additional twenty-five 727 passenger airliners, and signed options for

Galley servicing and fueling at major thru-stops do not interfere with passenger movements and cargo handling. Average ground time is eighteen minutes. Two easily accessible water connections are located within easy reach of the ground. Potable water and toilet facilities are designed to require minimum servicing.

The deep cargo compartments permit efficient cargo handling. Large doors are only forty-six inches above the ground to simplify loading and unloading by hand or cargo conveyor belt. So said the 1965 Sales Brochure for the 737.

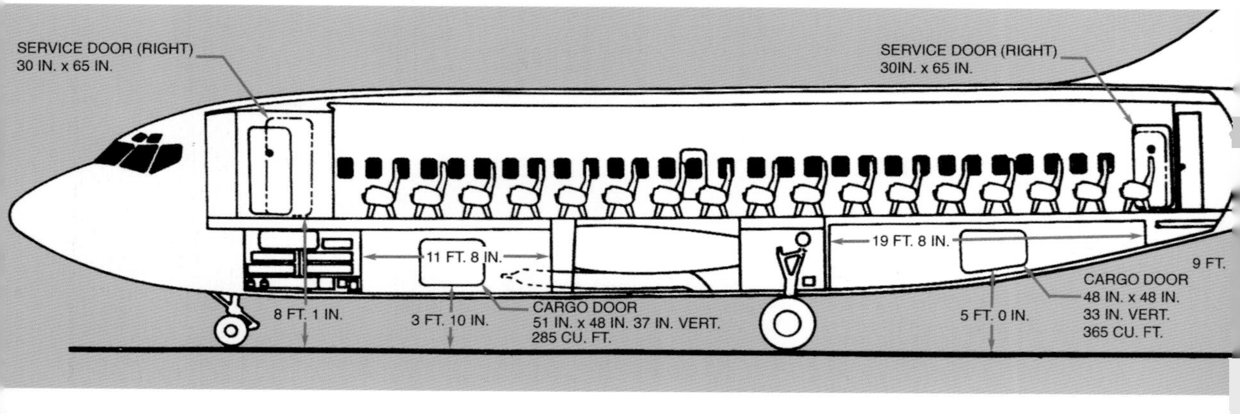

thirty more 737s and nine more 727QCs.

The first 271 737s were built in Seattle at Boeing Plant 2, just over the road from Boeing Field. However, with the sales of all Boeing models falling and large scale staff layoffs in 1969, it was decided to consolidate production of the 707, 727 and 737 at Renton, just five miles away. In December 1970 the first 737 built at Renton flew and all 737s have been assembled there ever since.

However, not all of the 737 is built at Renton. Since 1983 the fuselage, including the nose and tailcone has been built at Wichita and brought to Renton by train on custom-built railcars. They appear green due to their protective zinc-chromate coating. Much of the sub-assembly work has also been out-sourced.

Production methods evolved enormously since the first 737 was made in 1966. Instead of the aircraft being assembled in one spot, they are built on a moving assembly line similar to that used in car production. This accelerated production, not only reducing the order backlog and waiting times for customers but also reducing production costs. The line moves continuously at a rate of two inches per minute; stopping only for worker breaks, critical production issues or between shifts. Timelines painted on the floor help workers gauge the progress of manufacturing.

When the fuselage arrives at Renton, it is fitted with wiring looms, pneumatic and air-conditioning ducting and insulation before being lifted onto the moving assembly line. Next, the tailfin is lifted into place by an overhead crane and attached. Floor panels and possibly galleys - dependant on customer demands - are installed, and functional testing begins. In the 'high blow' test, the aircraft is pressurised to create a cabin differential pressure equivalent to an altitude of 93,000 feet. This ensures that there are minimal air leaks and that the structure is sound. Then the aircraft is jacked up for landing gear retraction and extension systems tests. As the aircraft moves closer

Simplicity keynotes the control cabin arrangement. The airplane is designed for two-pilot but can be flown by one pilot from either seat.

Instruments and controls are readily accessible to Captain and First Officer. Instruments in the 737 control cabin are either identical or similar to those in the 727 with the exception of approximately ten new instruments. Panel controls are arranged to provide a minimum workload with panel controls that must be operated most frequently located closest to the crew member.

A stowable Observers seat, installed in the isleway fold out of the right equipment shield. A further optional Observers seat can be installed behind the captain.

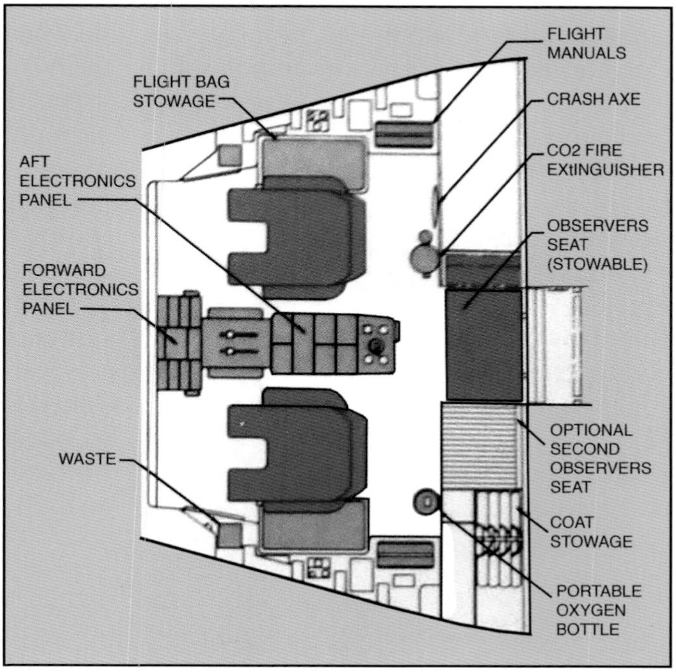

to the end of the line, sometimes the cabin interior is completed — seats, lavatories, luggage bins, ceiling panels, carpets... some airlines prefer to fit their own at their base. The final stage is to mount the engines. There are approximately 367,000 parts in a 737, held together by approximately 600,000 bolts and rivets.

After construction, they make one flight, over to Boeing Field, where they are painted and fitted out to customer specifications. It takes about fifty US gallons of paint to paint a 737, weighing over three hundred pounds per aircraft, depending on the livery. Any special modifcations or conversions are done at Wichita after final assembly of the green aircraft. Auxiliary fuel tanks, winglets, SATCOM and specialist interiors for Boeing Business Jets are fitted by PATS at Georgetown, Delaware as these are STC (Supplemental Type Certificate) items.

The two or three crew issue.

The 737 was the first two-crew aircraft produced by Boeing; all others had a flight engineer station which was necessary as early airliners had been more complex and less reliable. The three-crew issue had been around since the late 1950s when on 21 July 1958 a Presidential Emergency Board issued its report on a dispute between Eastern Air Lines and unions representing its pilots and flight engineers. President Eisenhower had appointed the Board the previous January to mediate the controversy over the qualifications of the flight engineer on turbojet transports. The Board concluded that a flight engineer on jetliners should have piloting qualifications and recommended that Eastern train its flight engineers to qualify for a commercial pilot's certificate. Despite the board's report in the Eastern dispute, American Airlines decided to give the third seat on the Boeing 707 to mechanic-trained flight engineers. Reacting to that decision, American's pilots walked off the job on 19 December. After 23 days, the strike ended when American agreed to add a third pilot (a fourth crew member) to the 707 flight deck. Other airlines that traditionally employed mechanic-trained flight engineers (Pan Am, Western, Eastern, and TWA) signed similar labour agreements with the Air Line Pilots Association requiring them to employ a fourth person in the jet cockpit.

When the 737 was announced with just two crew on the flight deck, the Airline Pilots Association (ALPA) and the Federal Aviation Authority (FAA) were on the case from the outset as the three-crew issue had still not been resolved.

Although getting somewhat ahead of the story, to finally determine if the aircraft was capable of safe two-crew operation, a 737 was flown with an FAA pilot and a Boeing pilot over Thanksgiving, the busiest week of the year in the Boston - Washington corridor. They flew forty sectors in six days, including approaches to minimums, go-arounds, diversions, simulated instrument failures and crew incapacitation. In December 1967 the FAA issued a statement declaring that '...the aircraft can be safely flown with a minimum of two pilots.'

Even after the FAA statement, American, United and Western continued to operate with three pilots until 1982. Fortunately, the rest of the world was not so limited and this helped sales to recover.

Chapter Two

The -100 Series

The decision to go ahead with the launch of the new design - which was expected to incur $150 million in development costs - was made by the Boeing board on 1 February 1965.

Just nineteen days later Lufthansa became the launch customer with an order for twenty-one aircraft, worth $67 million in 1965 dollars after the airline received assurances from Boeing that the 737 project would not be cancelled. Consultation with Lufthansa over the previous winter resulted in an offered increase in capacity to 100 seats.

On 5 April 1965, Boeing announced an order by United Airlines for forty 737s but the airline wanted a slightly larger airliner than the original 737, so Boeing stretched the fuselage thirty-six inches ahead of, and forty inches behind the wing. The longer version was designated 737-200, with the original short-body aircraft becoming the 737-100.

Engineers brought the sub-assemblies together, the fuselage joined with the wings and landing gear and then moved down the assembly line for the engines, avionics, and interiors. After rolling out the aircraft, Boeing tests the systems and engines before the airliner's maiden flight to Boeing Field, where it is painted and fine-tuned before delivery to the customer.

Though the prototype was first rolled out of the factory in September 1966, this prototype aircraft was introduced to the world on 17 January 1967 at Boeing's 'Thompson Site' located at Boeing Field. It was christened by

BOEING 737-100

Length:	94 ft 0 in (28.34 metres)	**Fuel (max):**	4,232 US gals (16,019 lit)
Height:	37 ft 0 in (11.21 metres)	**Range:**	2000 miles (3218 km)
Span:	93 ft 0 in (28.65 metres)	**Cruise Speed:**	570 mph (917 km/h)
Empty Weight:	58,607 lbs (28,580 kg)	**Pax Capacity:**	76-103
Max Weight:	110,000 lbs (49,885 kg)	**Powerplants:**	2 x P&W JT8D

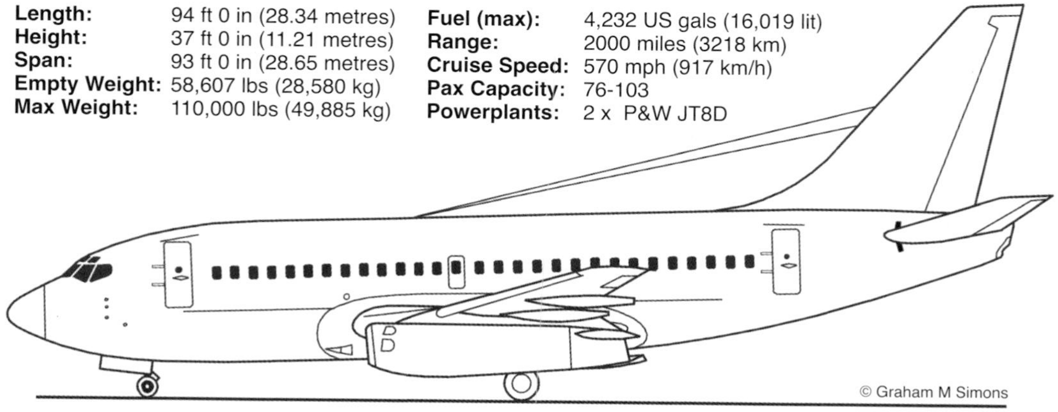

© Graham M Simons

uniformed flight attendants from the seventeen airlines which had placed 737 orders at the time.

The assembly building at Plant 2 wasn't tall enough for the 737's vertical stabiliser, so it had to be attached by crane in the plant's parking lot, then rolled down to the Thompson Site (where 392 additional 737s were built). This tail attachment process continued for the first eight 737s built.

The first aircraft was serial number 19437, registered N73700 and internally known as PA-099. It performed high-speed taxi tests at Boeing Field on 8 April 1967 and made its first flight with Brien Wygle and Lew Wallick at the controls at 1:15 pm the following day. The media reported an uneventful two and a half hour flight from Boeing Field to Paine Field, Wygle is supposed to have said 'I hate to quit, the airplane is a delight to fly.' Boeing's president at the time, Bill Allen, told reporters after the flight, 'We'll still be building this plane when I'm in an old-man's home.' He passed away in 1985, and the 737 legacy continued.

However, eyewitnesses 'who were there' said it happened very differently. The first flight was delayed by minor electrical system troubles - something that is very common to any new complex machine. There was a circuit breaker that would not stay set. After a delay, it was decided to go ahead without the snag being fixed. Take off was made to the north on Boeing Field runway 31L. What happened on the flight deck has never been revealed, but the airliner certainly did not complete a two and a half hour flight - instead, the crew flew a large pattern and landed back on 31L ten to fifteen minutes after take-off, taxiing back to the Boeing ramp. As far as eyewitnesses recall, there never were any plans to go to Paine Field and the prototype

The first 737 comes together at Boeing Field in 1966. *(author's collection)*

N73700, the original Boeing 737 prototype, is given a champagne christening during the 17 January1967 roll out event by flight attendants representing the aircraft's customers. *(author's collection)*

was based at Boeing Field for the entire flight test programme with occasional testing at Grant County airport and Edwards Air Force Base.

During the first twenty-five hours of testing - when the FAA allowed only minimum crew on board - there was a spin incident. The aircraft was equipped with an escape chute for bailout, and the crew had parachutes - although they did not have them on. The parachutes were stored in a temporary wooden rack just aft of the flight deck and adjacent to the escape chute. While stalls were being done, there was an asymmetrical deployment of the leading edge devices - the port side, I believe. The aircraft entered a spin, and recovery was affected by the asymmetric application of power. It was later reported that Brien Wygle had said that if he had been wearing his chute or had an ejection seat, he would have been gone'.

Flight testing continued at a blistering pace with the prototype clocking up 47 hours 37 minutes in the first month. Soon six aircraft, including the first -200, were on the flight test and certification programme. Between them they flew 1300 hours of flight tests. Many changes were made to the aircraft in this time, including trying inflatable main landing gear door seals, although these were soon changed to simple rubber strips.

However, the earliest 737s had some problems, including clamshell door thrust reversers taken from the 727 which didn't work properly, and a shimmy in the

This image has appeared in other publications captioned as 'The prototype 737 in flight' - something that I have always been unsure of, despite it being a stunning picture. What makes me think that it is an artwork, not a photograph is that records say that the prototype 737, was registered N73700, where this image shows it as N7370. *(author's collection)*

landing gear, '...but it was a good airplane from the start', recalled Brien Wygle. On 15 December 1967, the Federal Aviation Administration certified the 737-100 for commercial flight, issuing Type Certificate A16WE. The 737 was the first aircraft to have, as part of its initial certification, approval for Category II approaches, which refers to a precision instrument approach and landing with a decision height between 98 and 197 feet.

The first 737 went into service with Lufthansa on 10 February 1968. Generally, operators were very impressed with the reliability of the aircraft from the start, although inevitably there were some technical issues found during line work. The APU tended to shut down under load; this was solved by developing a new acceleration control thermostat. Engine starter valves were found to clog with sand from treated runways; this was fixed by the simple expedient of using a finer mesh on the filter screen.

In Lufthansa service, the Boeing 737 - called 'City Jets' by the airline - became the workhorse of the short-haul routes within the country and to other areas of northern Europe. Popular with passengers, the airliner was immediately accepted as a member of the family by crews who nicknamed it 'Bobby' much to the puzzlement of other nationalities. Why Bobby? The kids' book *Bobby Boeing Earns His Wings* was published in 1969 - the first in a whole series of titles produced by Lufthansa. In it the Boeing 707 was the father, the 727 the mother, and the 737 was the baby – called Bobby. The crews started using the name, and it stuck.

Boeing had always intended to make the aircraft easy to operate from airports that held little-to-no ground handling equipment, so the 737 came with built-in retractable airstairs for the port side passenger door while another was available as an option for the rear door. These proved to be troublesome, for the airstairs had a complicated way of folding and simultaneously collapsing the handrail as they retracted or extended. They also weighed 390 pounds, a significant penalty when viewed against potential load! Other minor issues were nosewheel corrosion, ram air inlet problems and hydraulic line failure, all of which were sorted out within a couple of years.

The 737-100 design schematics measured the aircraft at only ninety-four feet long, with a ninety-three foot wingspan, leading some to call it 'The Square Plane' or 'Fat Albert'. It carried 115 passengers and had a Maximum Take-Off Weight (MTOW) of 93,500 pounds, less than half that of the current -900 series built in 2019.

The original choice of powerplant was the Pratt & Whitney JT8D-1, which generated 14,000 pounds thrust, but by the time negotiations with Lufthansa had been completed the J T8D-7 was used. The -7 was rated to develop the same thrust at higher ambient temperatures than the -1 and became the standard powerplant for the -100.

N73700 never flew for an airline, though at one time it was painted in Lufthansa's livery. Both early models received their type certification on 15 December 1967. Throughout 1968, N73700 toured North and South America.

In February 1969, Boeing began offering a gravel kit on the 737 so it could operate from unpaved runways in remote locations. N73700 was used to test this kit, which included a deflection

Below: An air-to-air of what is assumed to be the prototype 737, although no registration is visible. The airliner is fitted with the short-length powerplants and carries the word 'Experimental' by the door.

Right: Brien Wygle (left) and Lew Wallick after the first test flight of N73700.
(both author's collection)

Above: Is it N73700 or N1359B showing off the gravel kit as it gets airborne from the Hope Airpark British Colombia during August 1972? My money is on the latter! *(Robin Banks collection)*

The photograph below does just show the registration N73700 by the rear door on the original. A comparison of the windows in the two pictures clearly does show they are totally different, and the above aircraft is a Series 200. *(author's collection)*

ski behind the nose gear, an oversized main gear, protective shields over the hydraulic tubing and brake cables on the main gears.

In 1972, N73700 toured Australia that March then performed further gravel testing in Peru that June. In August of that year, it performed grass field testing at Hope Airpark, British Columbia, about 100 miles north-northeast of Seattle. Lew Wallick, Boeing's Chief of Commercial flight test back in the 70s and 80s, used to give talks and showed movies of the testing at Hope. 'It was a 4500 foot grass runway used by light aircraft and gliders. We did both dry and wet grass landings, and take-offs and the aircraft behaved very well, although we had to call in tankers to spray the grass as it was very dry. After the test flights,

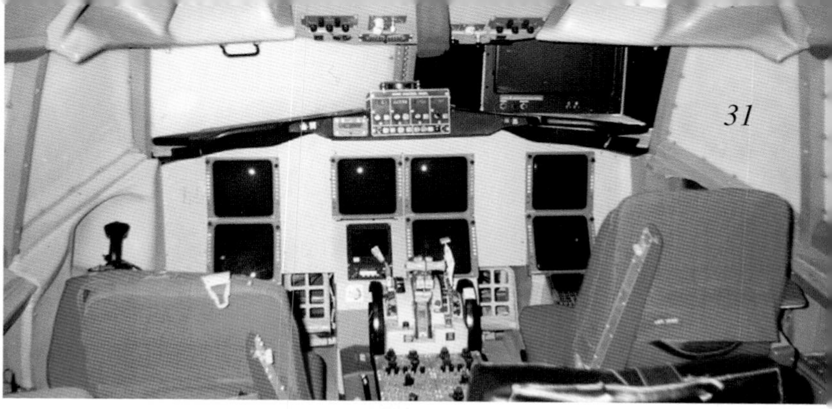

NASA 515 has two cockpits! One located where the standard cockpit would be, in the nose - the second placed where First Class would normally be located. Inside the second cockpit, it looks like the hideous love child of an Airbus and Boeing, with glass displays and side stick controls rather than the yoke common to Boeing jets.

Top: the 'internal' glass flightdeck with sidestick controllers that was installed where the First Class cabin would have been on NASA 515.

Right: the more conventional flightdeck with analogue instrumentation. The box on the coaming is the overide controls for the second flightdeck.

Below: NASA515 - also known as N515NA, formerly N73700, the first 737.
(all Simon Peters Collection)

Hope Airfield billed Boeing to have the runway re-graded to remove the ruts we had created with the 737!'

It is here that something of a mystery has arisen. Boeing, the Museum of Flight ... her ... ew ... to ... pe. ... ort

3 pilots for 737
NASA 515
+ 737 100 series
1st flight (33)

side of the airliner as it takes off in the video show sixteen windows between the main entry door and the escape hatch, proving that it could not be a Series 100, so it looks as if was actually N1359B, a B737-222 construction number 19758 - line number 16 that was used by Boeing Flight Test from 1970 to 1974. N73700 completed its career with Boeing in May 1973, with a tour to the Far East.

Above: Boeing 737-130 D-ABEF, the seventh 737 built on a publicity shoot. The aircraft is fitted with the short engine nacelles, that were later replaced.

Right: The 737 gained more nicknames than any other airliner, that included Tin Mouse, Maggot, Pocket Rocket, FLUF (Fat Little Ugly F**cker), Light Twin, Baby Boeing, Fat Freddy, Guppy, Thunder Guppy Yuppy Guppy, Super Guppy (series 3/4/500), Pig, Fat Albert... One of the more mysterious was Bobby, which came from a series of childrens cartoon books that included *Bobby Boeing's Abenteuer* (Bobby Boeing Adventure) produced by Lufthansa.

Below: the interior of a Lufthansa 737-130.
(all via Lufthansa)

First flight

of the world's newest short-range jetliner!

Boeing Field, Seattle. April 9. A gleaming new jetliner speeds down the runway and lifts quickly into the air.

The first flight of the Boeing 737—the world's newest, most advanced short-range jetliner—begins.

Like the first flights of earlier Boeing jets, the 737's was strictly routine, strictly as planned.

Like earlier Boeings, the 737 performed beyond even the highest expectations. During the 2½-hour first flight, the pilot radioed: "I hate to quit . . . this airplane is a delight to fly." Other flight crew reports: "Beautiful handling. Very nimble. Very responsive." Airline pilots are going to love the new Boeing 737.

So are passengers. Because the 737 Twinjet, the smallest member of the Boeing family, has a cabin as wide and roomy as the big 707s. It'll be the first short-haul jet to carry passengers between smaller cities in the wide-cabin comfort enjoyed by passengers flying between the capitals of the world.

The brilliant new Boeing 737 Twinjet is now undergoing intensive flight testing. Deliveries begin later this year. These airlines have already ordered 737s: Avianca, Braathens, Britannia, Canadian Pacific, Irish, Lake Central, Lufthansa, Mexicana, Nordair, Northern Consolidated, Pacific, PSA, Pacific Western, Piedmont, South African, United, Western, Wien Air Alaska.

BOEING 737

Above: another grey, dreary day at London Heathrow in 1971 as Lufthansa's 737-130 D-ABEP is seen parked on the stand at Terminal Two. *(Author)*

Right: The flightdeck of a Lufthansa 737-130, showing the plethora of 'round' dials and gauges. Note also the large undercarriage selector lever on the co-pilot's side of the panel! *(Lufthansa)*

The prototype Series 100 went to NASA on 12 June 1974, where it became known as the Transport Systems Research Vehicle (TSRV). It was in regular experimental use until 1997 and is now on loan from NASA to the Museum of Flight at Boeing Field, where it is on permanent display. NASA 515, as it was called, was involved in numerous pioneering flight investigations including control systems, 3D and 4D navigation, in-flight energy management, computerized flight management systems, electronic displays, Microwave Landing System (MLS) development (overrun in late development by the advent of GPS), slippery runway studies, and clear air turbulence and wind shear detection and warning. The airliner had a second flight deck, fully functional, installed in the main cabin, which was used for much of the flying. Control systems used included the original Boeing control column, 'Umbrella' handles (like bicycle

Lufthansas Fl Deck

ort-range jet et comfort.

That's the Boeing 737's big plus for passengers.

The new 737 is the smallest member of the Boeing family of jetliners. Yet it's as wide and roomy as the big 707 Intercontinentals. It's the only airliner that can bring passengers on short-range routes the same wide-cabin comfort travelers enjoy aboard transcontinental and over-ocean flights.

Equally important, the new 737 inherits all the benefits of Boeing's unmatched experience as builder of the world's most successful jetliners.

Newest, most advanced short-haul jet in the world, the 737 is currently undergoing an intensive flight test program. Pilots describe it as a delight to fly. The 737 takes off quickly, quietly. It cruises at 580 miles an hour, and carries up to 113 passengers.

The superb new Boeing 737 goes into service early next year. It has already been ordered by: Avianca, Braathens, Britannia, Canadian Pacific, Irish, Lake Central, Lufthansa, Malaysia-Singapore, Mexicana, NAC-New Zealand, Nordair, Northern Consolidated, PSA, Pacific, Pacific Western, Piedmont, South African, United, Western, Wien Air Alaska.

BOEING 737

Left: MSA 'a 737-112 9M-AOU seen on delivery. *(author's collection)*

Below: ZK-NEB of Ansett New Zealand, the former D-ABEC of Lufthansa. *(Simon Peters Collection)*

handlebars), and the current side-stick controllers. Numerous glass cockpit CRT display configurations were tried, that attacked many questions regarding display arrangements, colours and symbology. The aircraft had a unique fourth hydraulic system with a reservoir, pump and filter system. Other studies were performed of drag-reducing external coatings, flight deck displayed traffic information, take-off performance monitoring, and precision flare guidance during landing touchdown.

Thirty series 100s were built, with twenty-one going to Deutsche Lufthansa AG, five for Malaysia Airlines and two for Avianca.

Type Codes and Customer Support

Here is as good a place as any to explain the customer codes as applied to the Boeing type numbers. Boeing Commercial Airplanes allocated a set of unique, fixed customer codes used to denote the original customer for airframes produced as part of Boeing's 7X7 family of commercial aircraft between 1956, when they introduced of the 707, and 2016 when they announced that they would no longer apply customer codes to any aircraft produced after a certain point, which would lead to their designators being the 'generic' type for the model. The codes were removed from the type certificates for each model relying instead on just the production line number. Furthermore, customer codes would not be used for the 787 and 737MAX, and would not be applied to the 777X either.

Out 'in the field' Boeing had a well-established series of Parts Stores spread around the world; for example, a Boeing Parts Store was established in Brussels, Belgium which provided items for customers in Europe, the Middle ERast and Africa. This store, administered by SABENA, had an inventory of high-demand components for all models of Boeing range, and operated twenty-four hours a day, seven days a week. Linking in with these stores controlled by the Boeing Customer Support Department was a large team of Customer Support Representatives spread around the world. The list for the early 1970s show not only the main airports that were using the 737, but also the strength of staff at some of the locations.

Northern USA and Canada

***Seattle*	R W Wallace, Regional Director.
Anchorage	R E Reece.
Chicago	*M W Ludwig, D M Pettitt, E Huizinga.
Denver	H G Shea.
Edmonton	J P Mott.
Gander	M R Addis.
Kansas City	*N K Hillyer, C R Quinn.
Minneapolis-St Paul	*D G Holt, W A Mahan, M J Johnson.
Montreal	*D G Allen, M G Vogt
New York	*R A Nicholson, T M Martindale, R Niederkorn, B J Snyder, W A Staufenberg, J N Barber.
Sea-Tac	*R T Fellows, O E Lennon
Vancouver BC	W H McOsker, R P Aley.

Southern USA

***Los Angeles*	D V Cross, Regional Director. *M Cohen, J R Spear, L R Bott, R B Playstead, C H Whittlesey, J N Wasson.
Atlanta	*J P Thorsteinson, F Joyce.
Dallas	*F Habern, P Ross, L Shaw.
Honolulu	*W C Tattersall, R A Bogash.
San Diego	L F Hunt.
San Fransisco	*M L Blagsvedt, W R Chase, H A Rieke.
Tulsa	*D L Monchil.
Washington DC	N. E.Wolfe.
Winston-Salem	G S Anglin

Far East

Tokyo	M M Kiydno, Regional Director. *S R Harman, K F Mizuno, H L Bond, F W Piwenitzky.
Hong Kong	G W McElroy.
Seoul	*W J Campbell, W H Chappell.
Peking	*D J Cockerill, G L Totten, R C Smith.
Taipei	T K Tam.

Latin America

***Miami*	Connie Smith, Regional Director. E Bell, F E Hos, J E Matt, T Jones, W R Myers.
Bogota	W P Giesey.
Buenos Aires	A Cuevas.
Mexico City	H Vargas.
Porto Alegre	L J Sherry.
Rio De Janeiro	R E Prather.
Sao Paulo	D Stewart.

Europe

***Paris*	W S Thomas, Regional Director. * A J Mitchell, R E Franz
Amsterdam	*M Bodsky, E Quirie, J McCallum.
Belgrade	G J Wattier.
Brussels	*G R Nicoll, M P Hanvey.
Bucharest	* M Turner, W Parks, D Romine.
Copenhagen	R E Tuttle.
Dublin	B. M. Sorensen.
Frankfurt	*K Taht, R E Fairbanks.
Hamburg	*T F Barrett, J D. Rodrigues.
Kinshasa	R Friars.
Lagos	C A Von Thielmann.
Lisbon	*D S Kalotav, T C Montemayor.
London	*J W Purvis, D L Brewer, B D Allen, B E Hubbard.
Luton	L R Pestal.
Madrid	*T J Ellis, I J Jimenez.

PEOPLExpress' N408PE, the former 737-130 D-ABEO of Lufthansa. *(author's collection)*

Rome	*A Palatini.
Stavanger	R S Shafer.
Tel Aviv	W R Homer, D H Portman.
Zurich	F C Baker.
Australasia	
***Canberra*	C H Halbert, Regional Director.
Bangkok	W A Ulmstead.
Christchurch	E. Rose.
Kuala Lumpur	E. W. Pettitt.
Melbourne	L Lamb.
Singapore	*D B Erchinger, O Long.
Sydney	*A R Heitman, E W Berthiaume.
Middle East and Africa	
***Beirut*	Cliff Smith, Regional Director.
Algiers	R. L. Patterson
Amman	C W Riskedahl.
Athens	*C H Armstrong, L H Bennett, J Harp R A Russell, H A Sumner.
Bombay	A Bonham.
Cairo	H W Schuettke.
Casablanca	V C Rabbetts.
Jeddah	R. D. Bell.
Johannesburg	*K N Smith, M O Hansen.
Khartoum	K E Kells.
Kuwait	F Guthrie.
Lourenco Marques	I J Vogwill.
New Delhi	R. Fellows.
Tehran	H R Clark.
Tunis	C Gebara.

* - Senior Representative
** - Regional Headquarters

Lufthansa's early 737s remained in service for a remarkably long time, most surviving - albeit with different operators - well into the late 1990s. Apart from the three initial operators, Air Continente, Air California, Air Florida, AirCal, Aloha Airlines, America West Airlines, American Airlines, Ansett New Zealand, Asia Aviation Services, Avianca, Challenge Air International, Chapter Pty International, Condor, Continental Airlines, Copa Panama, Far Eastern Air Transport, Faucett, Magnicharters, Mexican Air Force, Oper Aereos, People Express, Presidential Air, SARO, Savar and Sierra Pacific Airlines all put their mark on the 737-100.

The last airworthy 737-100, line number 3, which first flew 12 June 1967, was finally retired from Aero Continente in Peru as OB-1745 in 2005.

Chapter Three

The -200 Series

Boeing immediately realised that most airlines - especially the potentially huge American market - wanted a slightly higher passenger load, to which they responded with the 737-200 of which there was eventually a myriad of sub-variants. Two sections were added to the fuselage; a thirty-six-inch section forward of the wing at Station 500A&B and a forty-inch section aft of the wing at Station 727A&B. As a result, the -200 version offered room for up to 115 passengers with a seat pitch of thirty-four inches or as many as 130 with a twenty-eight inch seat pitch.

All other dimensions remained the same. The JT8D was increased to 14,500 pounds thrust with the -9. Six weeks later, on the 5 April 1965, the -200 series was launched with an order for forty from United Airlines. Development and production of the two series ran simultaneously.

The first -200 series was 737-222 N9001U, the initial United aircraft of the order for forty, which made its maiden flight (as N7560V) on 8 August 1967.

By introducing the new model so early, indeed before the -100 prototype had even been built, Boeing was able to develop both aircraft simultaneously and incorporate any improvements in design into both models at an early stage, a process that helped to both reduce costs and improve the design and manufacturing process.

Once certification had been granted, United Air Lines worked up to introduce the aircraft into service, initially at the end of April 1968. The smaller -100 was soon discontinued, and Boeing concentrated its efforts on selling the -200, though with only gradual take-up by the airlines. More improvements were made, including allowing the use of twenty-five degrees of flap for take-off

BOEING 737-200

Length:	100 ft 0 in (30.48 metres)	**Fuel (max):**	4,232 US gals (16,019 lit)
Height:	37 ft 0 in (11.21 metres)	**Range:**	2200 miles (3540 km)
Span:	93 ft 0 in (28.34 metres)	**Cruise Speed:**	570 mph (917 km/h)
Empty Weight:	62,026 lbs (28,130 kg)	**Pax Capacity:**	115-130
Max Weight:	114,500 lbs (51,925 kg)	**Powerplants:**	2 x P&W JT8D

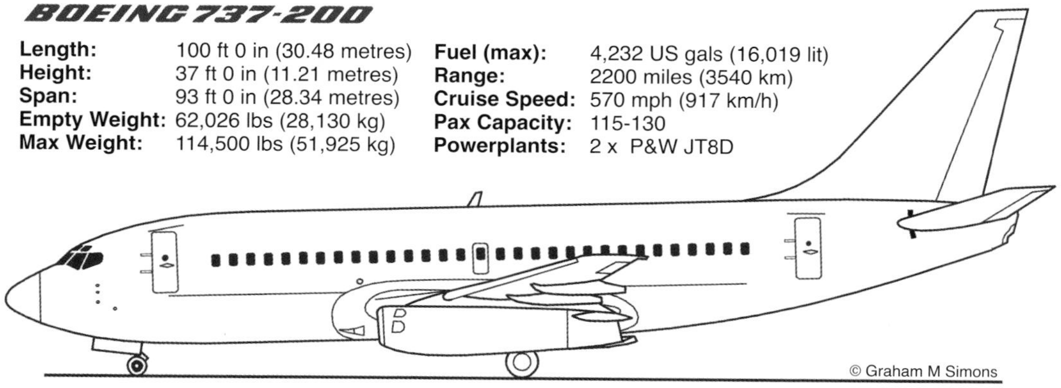

© Graham M Simons

An example of the early 737 production is G-AVRL, Britannia Airways' first 737 in the assembly hangar at Boeing Plant 2, Boeing Field. *(Boeing via Britannia Airways)*

and unrestricted use of forty degrees of flap on landing approach. These efforts, combined with the other aerodynamic modifications, helped to vastly improve the aircraft's performance - in particular, its short-field capability - during the early 1970s.

Flight testing had shown a five per cent increase in drag over predicted figures; this equated to a thirty-knot reduction in cruise airspeed. After almost a year of wind tunnel and flight testing, several aerodynamic modifications were made. Flaps and thrust reversers were improved from aircraft number 135 in March 1969 with free modification kits being made available for retrofit by existing customers. The thrust reversers were totally redesigned by Boeing and Rohr since the aircraft had inherited the same internal pneumatically powered clamshell thrust reversers as the 727, which were ineffective and apparently tended to lift the aircraft off the runway when deployed!

The redesign to external hydraulically powered target reversers cost Boeing $24 million but dramatically improved its short-field performance, which boosted sales to carriers proposing to use the aircraft as a regional jet from short runways. Drag reduction measures included extending the engine nacelles by three feet nine inches and widening the

Britannia Airways' G-AXNC departs the busy apron area at Luton, to the north of London, heading out 'up the hill' to the runway. *(author's collection)*

CR-BAA of DETA Lineas Aereas de Mocambuque was a 737-2B1, and first took to the skies on 4 December 1969. It survived until 2002 *(author's collection)*

strut fairings; enhanced flap, slat and panel seals; the elimination of thirteen inboard wing upper surface vortex generators and shrinking the size of the rest.

For many years United remained the only major US carrier to order large numbers of 737s because although the aircraft was designed to be flown by two crew, the American flight-crew trade union ruled that aircraft in that class had to be flown by three crew. United was forced to fly their 737s with three crew until 1981. Over in Europe, Air France had also been trying to order the aircraft for several years but was unwilling to do so because of staff opposition to two crew operation until after 1981.

With sales in the USA being damaged by the union requirement for a three-crew flight deck, Boeing decided to broaden its potential customer base by also looking beyond Europe to Africa and the Asia Pacific region. To assist this pursuit of new markets, Boeing was eager to improve further the aircraft's short-field performance, seen as a major selling point in countries that had many airports with short runways.

The original 737 series 200s - retrospectively named Series 200 Basic - had narrow engine pylons and smaller inboard edge Krueger flaps. The later series 200s had broader engine pylons and inboard leading edge Krueger flaps that extended to the fuselage.

From a transparency that has suffered colour degredation over the years is this image of a Western machines, N4525W. *(author's collection)*

Unpaved Strip Kit

This was made available for the 737-100/200 as early as February 1969. It allowed aircraft to operate from gravel, dirt or grass strips.

Whatever the surface, certain guidelines had to be observed. It had to be smooth with no bumps higher than three inches in one hundred feet or four inches in two hundred feet; good drainage with no standing water or ruts; and the surface material had to be at least six inches thick with no areas of deep loose gravel. Boeing offered a survey service to assess the suitability of potential strips. If a surface was not particularly hard, it could still be used by reducing tyre pressure down to a minimum of forty pounds per square inch.

Components included vortex dissipators to prevent vortices forming at the engine intakes which could cause gravel to be ingested by an engine. These consisted of a small forward projecting tube which blew pressure regulated engine bleed air at fifty-five pounds per square inch down and aft from nozzles at the tip to break up the vortices. The dissipators were operated by a solenoid held switch on the overhead panel which switched off on an 'aircraft squat' switch so that climb performance was not affected.

The nose-gear had a gravel deflector to keep gravel off the underbelly. Smaller deflectors on the oversized main gear to prevent damage to the flaps. There were protective metal shields over hydraulic tubing and brake cables on the main gear strut and over speed brake cables. Glass fibre reinforced the underside of the inboard flaps, and there was a metal edge band on the flap-to-body seals. Abrasion-resistant 'Teflon' based paint was applied on wing and fuselage under surfaces. The under-fuselage aerials were strengthened, and the anti-collision light was made retractable. Finally, screens in the wheel wells were fitted to protect components against damage.

The nose gear gravel deflector was made of corrosion-resistant steel and had a sheet metal leading edge to give it aerodynamic stability.

When the gear retracted, the deflector was hydraulically rotated around the underneath of the nose wheel before seating into the fairing at the front of the nose wheel well. The rotation was designed to maintain a nose-up attitude during transit. No extra crew action was required to use the deflector, and in the event of a manual gear extension, springs and rollers would position it correctly.

The ground clearance of this nose-gear unit was only three and a half inches. It was enough to allow for flat tyre clearance, but care had to be taken when crossing runway arrestor cables, particularly to avoid taxiing over the 'doughnuts' that supported any cables.

Sudan Airways ST-AFL was a 737-2J8C fitted with a Unpaved Strip Kit and is seen here passing through London Heathrow on delivery. The smaller pictures show the engine vortex dissipators (above) and the gravel deflector (right). *(all author)*

A close-up of the 737-100/200 wing set up in landing configuration with the leading edge slats extended and the triple-slotted flaps set to forty degrees. This particular machine, believed to be the first for United was clearly engaged in test flying, for it carries the word 'Experimental' by the forward door. It also has the early, 'short length' engines. *(author's collection)*

Series 200 Adv.

As well as incorporating all of the later - modifications, the Series 200Adv included major wing improvements such as new leading-edge flap sequencing, increase in droop of outboard slats, an extension of the inboard Krueger Flap. This was to produce a significant increase in lift and a reduction of take-off and approach speeds for a better short field performance or a Maximum Take-Off Weight (MTOW) increase of five thousand pounds. Autobrake, improved anti-skid, automatic speedbrake for Rejected Take-Offs, automatic performance reserve and even nose-brakes became available. Again, kits were available for existing operators of the -200. With the JT8D-15 at 15,500 pounds the MTOW was now up to 115,500 pounds and the Maximum Landing Weight at 107,000 pounds.

In 1973 when noise was becoming a factor, the nacelle was acoustically lined by Boeing and P&W swapped one fan stage for two compressor stages in the JT8D-17 while increasing thrust to 16,000 pounds. The JT8D got up to 17,400 pounds thrust on the -l7R.

Crew Training

The original 737 had been designed for route structures of relatively short stage lengths. A survey of the market determined that most commercial customers interested in this airliner would be operating from field lengths of more than 5000 feet.

Continuing improvements to the 737 design resulted in the Advanced 737-200 - sometimes referred to as the -200ADV - which featured lower take-off and landing speeds, shorter stopping distances and, as an option, the more powerful Pratt & Whitney JT8D-17 engines with a thrust rating of 16,000 pounds. Automatic brakes became standard equipment and assured immediate braking on touch-down with a pre-selected rate of deceleration through the landing roll-out. The engine thrust reversers had been redesigned to improve their capability. Leading-edge high lift devices and trailing-edge wing flap improvements provided more significant lift and reduced approach speeds by four to eight knots compared with the original 737.

The improved performance made the airliner attractive to airlines which previously were unable to operate jet aircraft because of limited runway lengths on their routes. Boeing 737-200ADVs began to be worked on unimproved runways 5000 feet in length, and were certificated for operations on runways only 4000 feet long.

For an airline making its first purchase of any Boeing airliner type, Boeing Commercial Airplane Company included in the Sales Contracts at no added cost a

A 737 climbs away from the airstip at Boardman in Washington State. Here pilots learned the short-field technique required to safely operate the machine into and out of some of the more challenging European airports. *(author's collection)*

The UK had a number of what became known as 'holiday airlines' which flew the 737 - this example being Gatwick and Manchester-based Air Europe, which operated a number of -200s. *(Air Europe via author)*

flight crew training programme. It consisted of ground school, cockpit procedure training, simulator training, transition flight training, and finally post-delivery route and line flying assistance and checks. The magnitude of the programme varied with the number of crews trained, but the course and the quality of training were standardised and resulted in fully qualified crew members.

Boeing flight crew training for the 737 was based at Boeing Field International in Seattle, Washington. The Flight Crew Training Department, however, took advantage of the favourable weather, lower population density and excellent runways and airfield control facilities at Grant County Airport, Moses Lake, Washington, about twenty minutes by 737 from Seattle. There, a 13,000-foot runway - as well as tower personnel well used to Boeing methods and airline flight training activities - made for landings and take-offs at higher frequencies than would have been possible at any other base. The result was more training per flight hour.

In the case of the Advanced 737, the greater capability of the aeroplane had opened to jet operations fields with runways of 5000-foot length or less. Pilots making the transition directly from propeller-driven aircraft to jet-powered airliners, therefore, would be called upon

N1359B, a 737-222, registered to Boeing and seen at Maher Air Force Base, where it was based from 1970-1974. It was later transferred to Air New Zealand as ZK-NAM. *(Tom Singfield collection)*

Another 'year of 1969 allumni' is JA8401, a 737-281 of All Nippon Airways. *(author's collection)*

5RMFA, a 737-2B2 of Air Madagascar, is often recorded as being marked as 5R-MFA, which clearly it is not. It is another 'first flyer' of 1969. *(author's collection)*

to carry out line operations on runways a third as long as that stretching in front of them at Moses Lake. The need to give them the confidence to meet such an operational environment was viewed as necessary by those involved in instructing them at the Boeing Flight Training Center.

Many of the airports which airlines intended to operate the 73 had runways between five and seven thousand feet in length - seemingly quite adequate for the new jet. It was complicated however by the poor condition of many of the runway surfaces - a number were quite bumpy - and often the airports only had a single, dual-direction runway that could impose a severe set of crosswind performance circumstances on operations.

Fortunately, Boeing already had a 'typical' field of such a length available on the sunny side of the Cascade Mountain range at Boardman, on land held under a long-term lease from the State of Oregon.

Boardman was used for US Navy flight training during World War Two. It was about twenty minutes flying time from Seattle and about ten minutes from Moses Lake. The runway was hard-surfaced, 138 feet wide by 4200 feet long. There was no control tower or approach aids at the strip. Situated on a plateau above the Columbia River, it was surrounded by sagebrush and was far from any community that might object to the sound of jet airliner operations.

The runway at Boardman had Boeing-developed markings painted at each end. There was a broad white band on each edge 200 feet long, which began 500 feet from the threshold. It was the touch-down aim point. The narrow marking ended 500 feet further down the runway and indicated where touch-downs could be made with enough runway remaining for a safe stop.

Flight training consisted of a series of landings and take-offs such as those typically used in 737 pilot training. Fullstop landings, missed approaches, touch-and-go landings, approaches and landings with one engine throttled back, rejected take-offs from V1 speed (the calculated 'go - no go' speed) and simulated engine failures at V1 speed with continued take-off, were all accomplished. Each pilot made eight landings.

By the time the new 737 pilots reached actual flight training, they had undergone ground school which included the material to prepare them for short-field training, and they had completed their simulator training. They were aware by now of the need for correct rotation technique where delayed or insufficient rotation could result in using too much runway. They knew that accurate and prompt response in case of engine failure at V1 meant adequate margins for stopping or continuing the take-off. They had been shown the importance of the correct approach path and approach speeds. They knew that too high an approach speed could waste available runway. Positive touch-downs and prompt use of brakes, spoilers, and reversers were pointed out to be essential.

They possessed the information, and they were about to get the experience!

Boeing 737 flight crew training always followed the same pattern. Each instructor pilot was assigned two airline pilots. He began working with them in the simulator. Actual flight training sessions were each four hours long and were staggered so that each instructor pilot and his two pilots-in-training flew at different times of the day on subsequent days - morning, afternoon or night. Each flight session was preceded by an hour's pre-flight briefing and followed by an hour's post-flight discussion.

When short-field training was scheduled - always as part of a daylight session - the sequence began with a briefing that covered the material discussed earlier in ground school.

Regarding take-offs, Boeing instructors reviewed engine pressure ratio (EPR) settings, go/stop decision execution, refused take-off transition and

On 6 December 1968, United, the second customer for the 737 after Lufthansa, received the one-hundredth 737 built. *(author's collection)*

implementation, and proper rotation which resulted in the earliest liftoff and best climb out. EPR settings are essential in short-field take-offs because settings even 0.10 low can result in 210 feet more runway being required to reach V1 and less altitude gained in a given distance. If the go/stop decision is delayed even a second, more distance is needed to stop after an engine failure than would be necessary if the decision were made and executed promptly.

Boeing instructed the pilots to set proper EPR for take-off; execute go/stop decision promptly because mis-execution of 'stop' could result in overrun and mis-execution of 'go' could result in reduced clearance height; rotate on schedule; rotate at a proper rate and rotate to appropriate attitude.

In preparation for the short-field landings, Boeing's instructors emphasised to their trainee pilots that the approach should be at the proper rate of descent and that the touchdown should be controlled, with the aircraft placed accurately on the runway and not allowed to float. They pointed out that while landing margins are large enough to allow for most additive errors, to ensure safe and comfortable operation, such errors should be minimised by the application of proper aircraft operation and cockpit discipline.

Series 200 Convertable (C) Quick Change (QC) and Combi.

Boeing was quick to offer airlines the option of a -200C convertible passenger/cargo variant to satisfy existing and future cargo requirements. The aircraft was equipped with an upward-opening cargo door on the port side of the forward fuselage to permit loading of standard seven foot four-inch by twelve feet five-inch cargo pallets of the same size as carried aboard the 707 and DC-8, the reason that Boeing chose the wider fuselage cross-section for the 737.

Rollerball flooring was installed to allow easy manoeuvring of the pallets within the fuselage. Typically, the -200C

Britannia Airways had a pair of 737-200QCs, this one is thought to be G-AXNA, seen here at Boeing Field, WA being prepared for its flight to a new home at Luton.
(Britannia Airways via author's collection).

C-FNVK, a 737-2R4C of First Air at start-up. *(Phil McCrackin Collection)*

could be configured to carry sixty-five passengers and three cargo pallets, but it could equally be converted to all-passenger or all-cargo in about an hour. In total, up to seven pallets could be accommodated, but only ninety-six model 200Cs were built, the first being handed over to Wien Consolidated on 30 October 1968.

Expanding even further on the concept, but designed to be even faster to convert, Boeing offered the -200QC Quick Change model in which even the seating was palletised. Each of the pallets held twelve seats and, together with palletised galleys, could be quickly exchanged for cargo pallets. The entire aircraft could be serviced and changed round in barely an hour, thanks to a special loader able to carry eight pallets at a time. However, the idea was not popular. Over the years, a number of -200 series aircraft were modified into a dedicated freighter configuration, which was regarded as a relatively cheap option for cargo operators to acquire a fast cargo aircraft for short to medium-haul routes.

A cabriolet 737

Aloha Airlines Flight 243 was a scheduled Aloha Airlines flight between Hilo and Honolulu in Hawaii and was being operated by Boeing 737-297 N73711, named *Queen Liliuokalani*. On 28 April 1988, the airliner suffered extensive damage after an explosive decompression in flight but was able to land safely at Kahului Airport on the island of Maui. There was one fatality, flight attendant Clarabelle Ho Lansing, who was ejected from the airliner. Another sixty-five passengers and crew were injured. Despite the substantial damage inflicted by the decompression and the loss of one cabin crew member, the safe landing of the aircraft established the incident as a significant event in the history of aviation, with far-reaching effects on aviation safety policies and procedures.

The captain of the flight was 44-year-old Robert 'Bob' Schornstheimer, an experienced pilot with 8,500 flight hours, 6,700 of which were on type. The first officer was 36-year-old Madeline 'Mimi' Tompkins; she also had significant experience flying the 737, having logged

Left: N464GB, a 737-293 called *'Mariner 1'* of Air California, seen landing sometime in the late 1960s. There is nothing unusual in that, until you study the picture below...

...taken in October 1968, not long after it was delivered. The rear galley service door appears to be the drop-down type affixed to the airstairs. Normally, the 737 had plug-type doors throughout. *(both author's collection)*

3,500 of her total 8,000 flight hours in that particular Boeing model. Some reports also say there was a Federal Aviation Administration air traffic controller on the flight deck as an observer.

Flight 243 departed from Hilo International Airport at 13:25 HST with six crew members and 89 passengers on board, bound for Honolulu. Nothing unusual was noted during the pre-departure inspection of the aircraft, which had already completed three round-trip flights from Honolulu to Hilo, Maui, and Kauai earlier that day, all uneventful. Meteorological conditions were checked but there were no advisories for weather phenomena reported along the air route, per AIRMETs or SIGMETs.

After a routine takeoff and ascent, the aircraft had reached its normal flight altitude of 24,000 feet when at around 13:48, about 23 nautical miles south-southeast of Kahului on the island of Maui, a small section on the left side of the roof ruptured with a 'whooshing' sound. The captain felt the aircraft roll to the left and right, and the controls went loose; the first officer noticed pieces of grey insulation floating above the cabin. The cockpit door had broken away and the captain could see 'blue sky where the first-class ceiling had been.'

The explosive decompression had torn off a large section of the roof, consisting of the entire top half of the aircraft skin extending from just behind the cockpit to the fore-wing area, a length of about eighteen and a half feet.

The captain took the controls, deployed the speed brakes and began an immediate descent at 280–290 knots with a rate of descent as high as 4,100 feet per minute. He turned toward the nearest airport, Kahalui Airport on the island of Maui. First Officer Tompkins handled all communications as well as assisting the

captain flying the airplane. Captain Schornstheimer described the flight controls as loose and sluggish.

There was one fatality: 58-year-old flight attendant Clarabelle 'C.B.' Ho Lansing, who was swept out of the airliner while standing near the fifth row seats; her body was never found. Lansing was a veteran flight attendant of 37 years at the time of the incident. Eight other people suffered serious injuries. All of the passengers had been seated and wearing their seat belts during the depressurization. In the very front of the airliner, attendant Jane Santo-Tomita was

As with the 707 and 727, the 737 was equipped with plug type passenger entry and service doors. The following pictures show the opening sequence. Boeing's instructions for opening the doors - and the illustrations - make interesting reading, showing an Aloha Airlines aircraft and Flight Attendant.

1. Standing on the left side of the door, grasp the handle and rotate in the direction of arrow well past the 'open' placard. 2. Grasp the assist handle while maintaining the same position on the floor while pushing outward on the door handle. 3 and 4. Rotate the door through the opening with as much force as possible to maintain your momentum.

The ill-fated N73711 of Aloha. *(Dr H Friedman Collection)*

hit violently in the head by debris and was slammed down to the floor. She was seriously injured but lucky to be down. A passenger held her there. Michelle Honda was also thrown to the floor as the plane started to decompress. She held on to the legs of the seats for her life.

The first officer tuned the transponder to emergency code 7700 and attempted to notify Honolulu Air Route Traffic Control Center (ARTCC) that the flight was diverting to Maui. Because of the cockpit noise level, she could not hear any radio transmissions, and she was not sure if the Honolulu ARTCC heard the communication. Although Honolulu ARTCC did not receive the first officer's initial communication, the controller working flight 243 observed an emergency code 7700 transponder return about 23 nautical miles south-southeast of the Kahalui Airport, Maui. Starting at 13:48:15, the controller attempted to communicate with the flight several times without success.

When the airliner descended through 14,000 feet, the first officer switched the radio to the Maui Tower frequency. At 13:48:35, she informed the tower of the rapid decompression, declared an emergency, and stated the need for emergency equipment. The local controller instructed flight 243 to change to the Maui Sector transponder code to identify the flight and indicate to surrounding air traffic control (ATC) facilities that the flight was being handled by the Maui ATC facility. The first officer changed the transponder as requested. At 13:50:58, the local controller requested the flight to switch frequencies to approach control because the flight was outside radar coverage for the local controller. Although the request was acknowledged, Flight 243 continued to transmit on the local controller frequency. At 13:53:44, the first officer informed the local controller, 'We're going to need assistance. We cannot communicate with the flight attendants. We'll need assistance for the passengers when we land.'

The captain stated that he began slowing the airliner as the flight approached 10,000 feet msl. He retracted the speed brakes, removed his oxygen mask, and began a gradual turn toward Maui's runway 02. At 210 knots IAS, the flightcrew could communicate verbally. Initially flaps 1 were selected, then flaps 5. When attempting to extend beyond flaps 5, the airplane became less

controllable, and the captain decided to return to flaps 5 for the landing. Because the captain found the airliner becoming less controllable below 170 knots he elected to use this speed for the approach and landing. Using the public address system and on-board interphone, the first officer attempted to communicate with the flight attendants; however, there was no response. At the command of the captain, the first officer lowered the landing gear at the normal point in the approach pattern. The main gear indicated down and locked; however, the nose gear position indicator light did not illuminate. Manual nose gear extension was selected, and still, the green indicator light did not illuminate; however, the red landing gear unsafe indicator light was not illuminated. After another manual attempt, the handle was placed down to

5 and 6: While maintaining the same position on the floor, continue to apply force to maintain momentum to the door. Some resistance will be encountered as the escape slide is pulled out of its compartment if the girt bar is installed. Continue to apply force and leverage without hesitation until the slide is pulled out of its compartment. 7 and 8: Continue pushing the door through its opening.

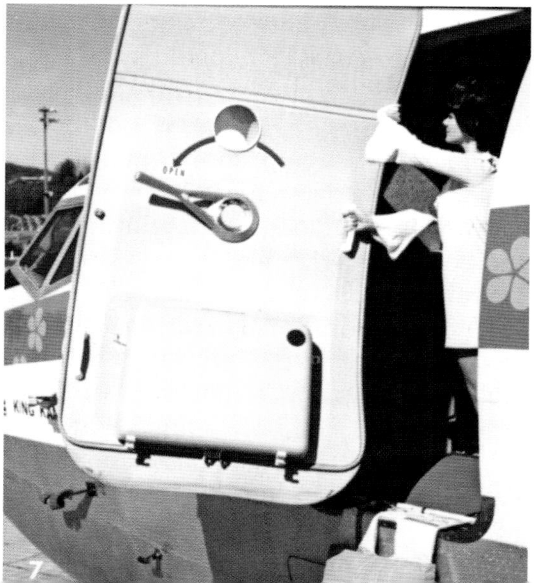

As the airliner came to a stop, this dramatic picture shows not only the damage but the passengers using the almost useless evacuation slide. *(FAA Collection)*

complete the manual gear extension procedure. The captain said no attempt was made to use the nose gear, downlock viewer, because the centre jumpseat was occupied and the captain believed it was urgent to land the airplane immediately. At 13:55:05, the first officer advised the tower, 'We won't have a nose gear,' and at 13:56:14, the crew advised the tower, 'We'll need all the equipment you've got.' While advancing the power levers to manoeuvre for the approach, the captain sensed a yawing motion and determined that the No.1 (left) engine had failed. At 170 to 200 knots IAS, he placed the No. 1 engine start switch to the 'flight' position in an attempt to start the engine; there was no response. A normal descent profile was established four miles out on the final approach. The captain said that the airplane was 'shaking a little, rocking slightly and felt springy.'

Flight 243 landed on runway 02 at Maui's Kahului Airport at 13:58:45. The captain said that he was able to make a normal touchdown and landing rollout. He used the No. 2 engine thrust reverser and brakes to stop the aircraft. During the latter part of the rollout, the flaps were extended to 40° as required for evacuation. An emergency evacuation was then accomplished on the runway.

A total of 65 people were reported injured, eight of them with serious injuries. At the time, Maui had no plan in place for an emergency of this type. The injured were taken to the hospital in tour vans belonging to Akamai Tours, driven by office personnel and mechanics, as the island only had two ambulances. Air traffic control radioed Akamai and requested as many of their fifteen-passenger vans as they could spare to go to the airport (three miles from their base) to transport the injured. Two of the Akamai drivers were former paramedics and established a triage on the runway.

Additional damage to the airliner included damaged and dented horizontal stabilisers, both of which had been struck by flying debris. Some of the metal debris had also struck the aircraft's vertical stabiliser, causing slight damage. The leading edges of both wings and both engine cowlings had also sustained damage. The piece of the fuselage blown off the aircraft was never found.

There were reports that one passenger, Gayle Yamamoto, told investigators that she had noticed a crack in the fuselage upon boarding, but did not notify anyone.

Aloha Airlines was a part of Boeing's Aging Fleet Program and was visited by the Boeing team for inspection of their highest flight/cycle time aircraft, several of which had exceeded seventy-five per cent of the 737's design life. Boeing personnel reported concern about the corrosion and skin patches they'd found on some of the fleet. Boeing recommended that Aloha totally strip and upgrade the structures, which would mean the aircraft being out of service for a month or two. Aloha Airlines hired a

9 and 10: The door will moves smoothly through the opening to the full open and locked position. During the entire sequence, the operator maintain her position at ther aft side of the entry way.

To close the door - 11 and 12 - assume a position at the aft side of the entry way, depress lock release on upper hinge nd grasp the assist handle. While keeping the arm straight, lean back into the cabin using body weight to swing the door closed. Rotate handle to the locked position to secure the door.

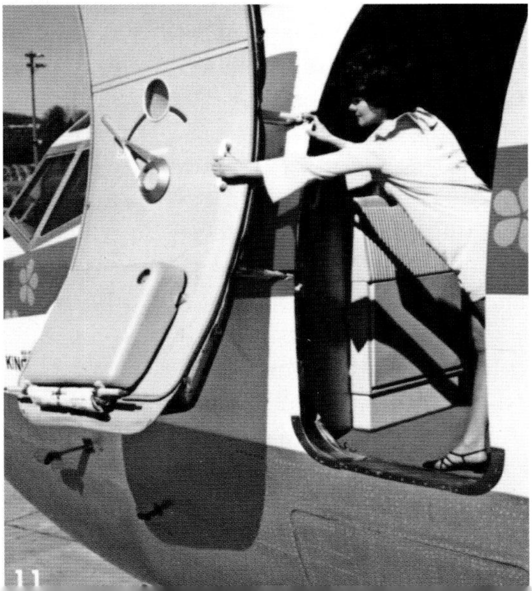

A poor quality, but interesting picture of N73711 after the drama was over. The picture shows not only the main fuselage damage but also the damage to the wings and tail. *(FAA Collection)*

number of new staff to deal with keeping their fleet safe, including a Staff Vice President of Quality Assurance and Engineering, a Director of Quality Assurance and a Chief Inspector. What they failed to do was prioritise stripping and upgrading the 737s as a matter of urgency. After all, their 737's flew short distances and at lower altitude, so the flight hours were nowhere near maximum and, although the cycles were high, the short flights often did not reach the maximum pressure differential, so the number of full pressurisation cycles were considered to be significantly fewer than the 75,000 the aircraft should be able to handle.

Aloha Airlines also pointed out that they were assured by Boeing that the aircraft was still safe to operate. Boeing's report with the recommendations was presented six months after the initial visits, clearly not giving the impression that they were urgent. Aloha Airlines felt confident that their aircraft were not suffering the same wear-and-tear as that of other operators and that there was plenty of time to get them upgraded.

The investigation by the US National Transportation Safety Board revealed that Aloha Flight 243 was the 152nd Boeing 737 airframe. It was built in 1969 and delivered to Aloha Airlines as a new aircraft. While the airframe had accumulated 35,496 flight hours before the accident, those hours included over 89,680 flight cycles, owing to its use on short flights. This amounted to more than twice the number of flight cycles it allowed for in its design.

For the aircraft production number 292 (B-737) and after (this aircraft was production line number 152), an additional outer layer of skin or doubler sheet at the lap joint of the fuselage was added. In the construction of this aircraft, this doubler sheet was not used in addition to other details of the bonding of the parts. In the case of production line 292 and after, this doubler sheet gave an additional thickness of 0.036 inches at the lap joint. In aircraft line number 291 and before, cold bonding had been utilised, with fasteners used to maintain surface contact in the joint, allowing bonding adhesive to transfer load within the joint. This cold bonded joint used an epoxy-impregnated woven scrim cloth to join the edges of 0.036 inch thick skin panels. These epoxy cloths were reactive at room temperature, so they were stored at dry ice temperatures until used in manufacture. The bond cured at room temperature after assembly. The cold bonding process reduced the overall weight and manufacturing cost. Fuselage hoop loads (circumferential loads within the skins due to pressurisation of the cabin) were intended to be transferred through the bonded joint, rather than through the rivets, allowing the use of lighter, thinner fuselage skin panels with no degradation in fatigue life.

The investigation determined that the quality of inspection and maintenance programmes was deficient. As fuselage examinations were scheduled during the night, this made it more difficult to carry out an adequate inspection of the aircraft's outer skin.

Also, the fuselage failure initiated in the lap joint along S-10L; the failure mechanism was a result of multiple site fatigue cracking of the skin adjacent to rivet holes along the lap joint upper rivet row and tear strap disbond, which negated the fail-safe characteristics of the fuselage. Finally, the fatigue cracking initiated from the knife edge associated with the countersunk lap joint rivet holes; the knife edge concentrated stresses that were transferred through the rivets because of lap joint disbonding.

The NTSB concluded in its 264-page final report on the accident: 'The National Transportation Safety Board determines that the probable cause of this accident was the failure of the Aloha Airlines maintenance program to detect the presence of significant disbonding and fatigue damage which ultimately led to failure of the lap joint at S-10L and the separation of the fuselage upper lobe. Contributing to the accident were the failure of Aloha Airlines management to supervise properly its maintenance force; the failure of the FAA to require Airworthiness Directive 87-21-08 inspection of all the lap joints proposed by Boeing Alert Service Bulletin SB 737-53A1039; and the lack of a complete terminating action - neither generated by Boeing nor required by

Left: Flight Attendant Clarabelle 'C.B.' Ho Lansing, who lost her life in the incident.

Above: Captain Robert 'Bob' Schornstheimer.

Right: First Officer Madeline 'Mimi' Tompkin.

In 1997 British Airways introduced its tails of the world' theme this one being called 'Benehone' as seen on 737-236 G-BGDL. The schemes were not liked by staff, and this design was nickednamed 'travel-rug'. *(author)*

the FAA - after the discovery of early production difficulties in the B-737 cold bond lap joint which resulted in low bond durability, corrosion, and premature fatigue cracking.

One board member dissented, arguing that the fatigue cracking was clearly the probable cause, but that Aloha Airlines maintenance should not be singled out because failures by the FAA, Boeing, and Aloha Airlines Maintenance each were contributing factors to the disaster.

Pressure vessel engineer Matt Austin has proposed an additional hypothesis to explain the scale of the damage to Flight 243. This explanation postulates that initially the fuselage failed as intended and opened a ten-inch square vent. As the cabin air escaped at over 700 mph, flight attendant Lansing became wedged in the vent instead of being immediately thrown clear of the aircraft. The blockage would have immediately created a pressure spike in the escaping air, producing a fluid hammer (or 'water hammer') effect, which tore the jet apart. The NTSB recognised this hypothesis, but the board did not share the conclusion. Former NTSB investigator Brian Richardson, who led the NTSB study of Flight 243, believed the fluid hammer explanation deserves further study.

737-200 Executive

An executive or business jet version of the Series 200 - originally designated the 737-32, and later the Corporate 200 - was also available, which was basically a standard 737-200 or the Advanced -200 but with a wide variety of interior configurations designed to suit the customer. Boeing was happy to fit the aircraft out but some customers preferred the aircraft 'green' for fitting out by a specialist company. Also available with this model was additional fuel capacity in the shape of fuel tanks located in the cargo hold.

A combination of factors limited sales of the 737 during most of the 1970s; the global oil crisis was one but the need for three flight deck crew in the USA certainly limited sales to the domestic market. However, towards the end of the decade sales boomed, especially after the passing of the 1978 Airline Deregulation Act and in that year Boeing sold 145 737s. This rate of sale continued into the 1980s, and in 1982 the 1,000th example was ordered. The final -200 model was 737-25C B-2524 (c/n 24236), which was delivered to Chinese carrier Xiamen Airlines on 2 August 1988.

Such was the popularity of the -200 that its production continued for over four years after the introduction of the first -300. One thousand one hundred and forty-four originals were built, some of which are still flying today, although noise restrictions have made necessary the installation of Stage Three hush kits available from Nordam or AvAero. P&W was even considering a re-engining programme with the PW6000, but with the post-September 11th economic downturn sending thousands of aircraft into storage, the idea was dropped.

Third-party companies were still developing the 737-200. In June 2005 Quiet Wing gained FAA certification for a flap modification package to increase take-off performance by 7,000 pounds, reduce fuel consumption by three per cent and reduce stalling speeds by five knots. It worked by drooping the trailing edge flaps by four degrees and the ailerons by one degree to increase the camber of the wing. Whilst this slightly increases drag, it does give much more lift, thereby increasing the aerodynamic efficiency of the wing. One unusual benefit was that operators wanted to replace their JT8D-15 engines with the older lower thrust but lower fuel consumption -9As.

'Dammit Janet....'

It is impossible to leave the 737-200 without making mention of the world's most famous secret airline 'Janet'.

Sometimes called Janet Airlines, it is the unofficial name given to a highly classified fleet of passenger aircraft operated for the United States Department of the Air Force as an employee shuttle to transport military and contractor employees. The purpose is to pick up the employees at their home airport and take them to their place of work. Then, in the afternoon, they take the employees back to their home airports. The airline mainly serves the Nevada National Security Site (most notably Area 51 and the Tonopah Test Range), from a private terminal at Las Vegas's McCarran International Airport. The airline's aircraft are generally unmarked but do have a red paint stripe along the windows of the aircraft, which

A Janet lands back at McCarren Airport, Las Vegas - from where, who knows? *(author's collection)*

Some colour schemes just look right. CP Air CF-CPU in flight *(CP Air via authors collection)*

gives some sort of hint at Janet being the operator.

The fleet's 'Janet'call sign, from which its de facto name comes is said to stand for 'Just Another Non-Existent Terminal'. It is also sometimes known as 'Joint Air Network for Employee Transportation'. Due to the airline's secretive nature, little is known about its organisation. It is operated for the USAF by infrastructure and defence contractor AECOM through AECOM's acquisition in 2014 of URS Corporation, which acquired EG&G Technical Services in 2002, as derived from URS's history of providing this service to the Air Force and job openings published by URS. For example, in 2010, URS announced it would be hiring Boeing 737 flight attendants to be based in Las Vegas, requiring applicants to undergo a Single Scope Background Investigation in order to obtain a Top Secret security clearance.

Due to its secrecy, Janet boarded at a special part of McCarran International Airport. Passengers board aircraft at the west side of the airport, next to the Janet Airlines passenger parking lot. There is even a small dedicated terminal building for them.

Janet flights operated with a three-digit flight number and a ICAO WWW-prefix. which, in the official publication of airline codes, this specific three-letter designator was listed as blocked. The official airline callsign is simply Janet. However, the airline also uses different callsigns, called Groom Callsigns once transferred over to Groom Lake from Nellis control. The callsign name would change, and the callsign number will be the last two digits of the flight number +15.

The first flights from Las Vegas to Area 51 were performed in 1972 by a Douglas DC-6 operated by EG&G. A second Douglas DC-6 was added in 1976, and this type remained in use until 1981. Janet operated a number of Boeing 737-200s - early examples bring former Western Airliners machines, some of which were modified from military T-43A aircraft. One of the 737-200s with registration N5177C in the 1980s was briefly based in Germany at Frankfurt International Airport (which was at the time also home to USAF Rhein-Main), and operated by Keyway Air Transport, apparently a front company for a US government operation.

T-43A, CT-43A, NT-43A

The T-43A - informally known as 'Gator', an abbreviation of the word 'navigator' after its duties - was a US Air Force version of the Boeing 737-200. The exterior differences between the military

A T-43A above Mount Rainier on it's pre-delivery photo-shoot. *(Boeing via author's collection)*

and commercial aircraft were that they only had nine windows on each side of the fuselage and door 1R and 2L were not fitted. There were also many small blade-type antennas for UHF communications and five overhead sextant ports. The aircraft was fitted with an 800 US gallon auxiliary aft centre tank as standard.

First flown on 10 April 1973, the first T-43A was delivered to the Air Education and Training Command at Mather AFB, Ca, in September 1973. The fleet relocated to Randolph AFB, Texas, in May 1993 when Mather closed.

Most of the T-43As were used in the USAF's undergraduate navigator training programme to train navigators for strategic and tactical aircraft. The rest were configured for passengers as the CT-43A and were assigned to the Air National Guard at Buckley, where they were used for the USAF Academy's airmanship programme and to provide travel service for academy sports teams. Also, US Southern Command had a CT-43 for commander transport.

Inside each T-43A training compartment were two minimum proficiency, two maximum proficiency and twelve student stations. Two stations formed a console, and instructors could move their seats to the consoles and sit beside students for individual instruction. The cabin floor was strengthened to take the weight of

T-43As in long-term storage at Davis-Monthan. *(Dr H Friedman collection)*

these consoles. The large cabin allowed easy access to seating and storage yet reduced the distance between student stations and instructor positions.

The student training compartment was equipped with avionics identical to that of Air Force operational aircraft. This includes Doppler and mapping radar; LORAN, VOR, TACAN, INS, radar altimeter; UHF and VHF Comms. Five periscopic sextants were spaced along the length of the training compartment for celestial navigation training.

Gradually most of the T-43s went into storage at the Aircraft Maintenance And Reutilisation Center (AMARC) at Davis Monthan AFB near Tucson, Arizona. However, one (tail number 73-1155) was recovered in March 2000 and flown to an aircraft maintenance and modification facility at Goodyear, Arizona for conversion to a radar testbed and became the NT-43A.

Throughout its service in the Air

Below: the sole NT-43A on final approach.

Right: 'Rat 55' is allegedly the call sign used by pilots flying this highly modified aircraft that is based at 'Area 51' in central Nevada. The word 'Rat' comes from the aircraft's function as a Radar Testbed and the '55' comes from the machine's USAF serial number 73-1155.

The NT-43A has been photographed tailing a B-2 stealth bomber over Death Valley and over the Tonopah Test Range in Nevada. The patch depicts a rat holding a radar in its right hand and another radar dish strapped to its rear-end, both of which recall the radome configuration on the NT-43A. The rat's hat recalls the wizard figures associated with classified flight test operations in other patches. *(author's collection)*

Training Command and the successor, Air Education and Training Command, no T-43 was ever lost in a mishap. However, among the T-43s removed from navigator training and converted to CT-43A executive transports, one aircraft assigned to the 86th Airlift Wing (86 AW) at Ramstein Air Base, Germany to support United States European Command (USEUCOM) crashed in Croatia in 1996 while carrying then-U.S. Secretary of Commerce Ron Brown and thirty-four other passengers. There were no survivors and the subsequent investigation determined that this was a controlled flight into terrain (CFIT) mishap as a result of pilot error.

Three 737-200s were modified as Maritime Patrol aircraft for the Indonesian Air Force, first flying on 21 April 1982 and named the MP Surveiller. Fitted with Motorola AN/APS-135(V) Side Looking Airborne Modular Multi-Mission Radar (SLAMMR), the antenna of which was mounted in two sixteen foot long housings on the upper rear fiuselage, either side of the vertical fin. This system could spot small ships at ranges of one hundred nautical miles. At least one Surveiller, AI-7303, was still in service with the Indonesian Air Force in August 2017.

Series 200 testbeds

The aforementioned NT-43A had two oversized radomes on the nose and tail. Its first flight in this new configuration was on 21 March 2001. The radomes were built by the Lockheed Martin Advanced Prototype Center, part of the Advanced Development Programs' (ADP) organisation for Denmar, a company specialising in stealth technology. The 'Den' in the name stood for President Denys Overholser, the former Skunk Works engineer credited with devising the shape of the first stealth aircraft. The design, fabrication and machining of the structure's components were all performed at Palmdale. The radome structure was 6.2 feet in diameter and 16.5 feet in length and made of a ninety per cent carbon epoxy honeycomb sandwich material, with machined aluminium parts.

The NT-43A was observed flying in formation with various stealth aircraft, usually in the radar free environment of Death Valley. It is believed that its task was to make radar images of these aircraft to evaluate their long-term stealth characteristics. The images could then be used to reveal the rate of degradation of the radar detecting and absorbing components as the aircraft age, and to

4X-AOT the Boeing 737-297 of IAI ELTA Electronics at the 2003 Paris Airshow. The aircraft had a series of interchangable nosecones to suit different mission requirements. *(author's collection)*

N737BG, Boeing's own much-modified 737-247 Avionics Flying Laboratory used for testing elements of the Joint Strike Fighter. It was the former Western Airlines N4515W. *(author's collection)*

determine the effectiveness of maintenance and repair methods.

The Isreali Aircraft Industries Elta Radar Testbed using a 737-200 airframe has been used to develop systems since 1979. These have included maritime patrol signal intelligence, image intelligence using synthetic aperture radar, AEW and most recently Flight Guard, a commercial aircraft anti-missile protection system.

Flight Guard was based around six miniaturised pulse-Doppler sensors, located to give all-round coverage. On the Boeing 737 demonstrator two are fitted below the nose radome, with two more further aft on the forward fuselage and two on the tailcone. These are used to automatically trigger the release of IR decoy flares in the event of an attack. The system gives greater than 99% probability of missile detection and has a very low false alarm rate.

Flare dispensers would usually be fitted in the wing-fuselage fairing as this does not involve penetrating the aircraft's pressure hull, and gives a minimum drag configuration. The system is armed or disarmed at the Flightdeck Control Display Unit.

The Boeing Avionics Flying Laboratory first flew as the AFL on 26 March 1999. However, this highly modified 737-200 had initially been flown in 1968. It was modified by Boeing Aerospace Support to accommodate special avionics and instrumentation for development of the F-35 Joint Strike Fighter.

The reason for the AFL according to Dan Cossano, manager of Boeing JSF Mission Systems: 'We will save development time and costs because the AFL allows us to test more efficiently than with a fighter platform'.

The design and modification programme took less than a year. Boeing Military Programs fitted a forty-eight-inch nose and radome assembly to the forward pressure bulkhead of the aircraft. The radome housed the JSF synthetic aperture radar and forward-looking infrared sensors used for targeting. The aircraft also was fitted with a JSF cockpit in the cabin, several antennas, a heat exchanger on the port side of the fuselage and provisions for a supplemental power system.

Chapter Four

Rivals from Europe

Airbus Industrie began as a consortium of European aviation firms formed to compete with American companies such as Boeing, McDonnell Douglas, and Lockheed.

Hardly noticed on the world stage, Airbus had begun in 1967 as a marriage of four European aircraft makers in Britain, Germany, France, and Spain.

No one knew if the aircraft manufacturer would ever turn a profit, as the Americans saw it, that was not the primary goal. The USA saw Europe as welfare-minded, and the idea behind Airbus was to save the European airframe business and the thousands of jobs that it would provide.

In addition, it was hoped that a European airliner builder could give national airlines - known as flag carriers - some relief from Boeing's vice-like grip on the top of the market, specifically the 747, which allowed monopoly pricing by Boeing. It was the same year that the first Concorde rolled out of the factory in Toulouse.

Both events were a part of the 'call to arms' and the embodiment of a new intensity by the Europeans to achieve dominance in technological advances.

By the mid-1960s, several European aircraft manufacturers had drawn up competitive designs but were aware of the risks of such a project. The European industry began to accept, along with their governments, that collaboration was required to develop such an aircraft and to compete with the more powerful US manufacturers. Negotiations began over a European collaborative approach, and at the 1965 Paris Air Show, the major European airlines informally discussed their requirements for a new 'Airbus' capable of transporting 100 or more passengers over short to medium distances at a low cost. The same year Hawker Siddeley - at the urging of the UK government - teamed with Breguet and Nord to study Airbus designs. The Hawker Siddeley/Breguet/Nord group's HBN 100 became the basis for the continuation of the project. By 1966 the partners were Sud Aviation, later Aérospatiale (France), Arbeitsgemeinschaft Airbus, later Deutsche Airbus (West Germany) and Hawker Siddeley (UK). A request for funding was made to the three governments in October 1966, but it was not until 25 July 1967 that the governments agreed to proceed with the proposal.

In the two years following this agreement, both the British and French governments expressed doubts about the project. The memorandum of understanding had stated that seventy-five orders must be achieved by 31 July 1968. The French government threatened to withdraw from the project due to its concern over funding all of the Airbus A300, Concorde and the Dassault Mercure concurrently, but was persuaded to maintain its support. With its own concerns at the A300B proposal in December 1968, and fearing it would not recoup its investment due to lack of sales, the British government withdrew on 10 April 1969. West Germany took this opportunity to increase its share of the project to 50%. Given the participation by Hawker Siddeley up to that point, France and West Germany were reluctant to take over its wing design. Thus the British company was allowed to continue as a privileged subcontractor.

Airbus Industrie was formally established as a Groupement d'Intérêt Économique (Economic Interest Group or GIE) on 18 December 1970. It had been formed by a 1967 government initiative between France, West Germany and the UK. Its initial shareholders were the French company Aérospatiale and the West German company Deutsche Airbus, each owning a fifty per cent share. The name 'Airbus' was taken from a non-proprietary term used by the airline industry in the 1960s to refer to a commercial aircraft of a certain size and range, as it was linguistically-acceptable to the French. Aérospatiale and Deutsche Airbus each took a 36.5 per cent share of production work, Hawker Siddeley twenty per cent and the Dutch company Fokker-VFW seven per cent Each company would deliver its sections as fully equipped, ready-to-fly items. In October 1971 the Spanish company CASA acquired a 4.2 per cent share of Airbus Industrie, with Aérospatiale and Deutsche Airbus reducing their total stakes to 47.9 per cent. In January 1979 British Aerospace, which had absorbed Hawker Siddeley in 1977, acquired a twenty per cent share of Airbus Industrie. The majority shareholders reduced their shares to 37.9 per cent, while CASA retained its 4.2 per cent.

The consortium's ownership evolved to two private corporations - eighty per cent for the European Aeronautic Defense and Space Company (EADS), and twenty per cent by British Aerospace Ltd, later BAe Systems. EADS officially emerged on 10 July 2000 as a result of the merger of Aerospatiale Matra, Daimler Chrysler Aerospace (DASA) and Construcciones Aeronauticas (CASA).

The newly incorporated EADS became the world's third largest aerospace group, behind Boeing and Lockheed Martin, with an estimated $22 billion in annual revenues and 96,000 employees.

1973 saw Airbus flying its first product, the A300. The airliner was technically as good as promised, but sales were dismal. To bolster interest, Airbus staged a round the world demonstration flight.

The threat of European competition

The 1974 SBAC Farnborough Airshow was wet and windy, but Airbus and Air France put on a sparkling display with the A300B2 F-WUAA *(author)*

had never been taken seriously at Boeing. Europeans had made brief entries into the USA but soon disappeared. Concorde failed to make any sales in the US after much political lobbying, spying and outright lies by politicians and those with vested interest (see *Concorde Conspiracy,* The History Press 2012, by the same author). Rumours abounded - totally unproven as it turned out - that there was widespread concern about the cohesiveness and after-sales support from a factory that straddled several national frontiers. Only eleven firm orders for the A300 were booked by the end of 1974.

Frank Frederick Borman II (*b*.March 14, 1928) Colonel USAF (ret) aeronautical engineer, test pilot, businessman, rancher, and NASA astronaut. He was also President and CEO of Eastern Airlines.

Initially, Boeing regarded Airbus Industrie - as it was initially known - as an upstart that would never be a threat with their A300. However, Airbus increased their efforts and began to offer attractive financial terms. Low interest rates and no principal payments for the first year were not uncommon. The terms were backed by the host governments.

Its perceived market niche - the large European flag carriers - was abandoned as too limited, and instead, they began seeking new customers around the world - and zeroing in on the US market.

Airbus fortunes abruptly changed in 1976, when they concluded a very lucrative deal with Eastern Airlines.

If ever there was an opportunity to head Airbus 'off at the pass' it was here, but Boeing dropped the ball, allowing President Frank Borman of Eastern Airlines to purchase twenty-three A300s with options on a further nine. The Eastern decision had not only wedged the door open to the American market, it served notice to the rest of the world that Airbus had arrived. Orders for the A300 jumped to sixty-nine airliners in 1978 and took off with ninety-eight in the first five months of 1979.

When Airbus Industrie announced a directly competitive airliner to the new technology Boeing 767 in mid-1978, the 224 passenger A310, the last of the Boeing doubters still had not disappeared, arguing that it was simply a shortened A300.

Robert Lloyd 'Bob' Crandall *b*. 6 December1935) the former President and Chairman of American Airlines.

The world's airlines now viewed European technology as equivalent to the Americans, and there was no match in the financial arena.

Airbus made its next great break into the US market when President Robert 'Bob' Crandall of American Airlines purchased thirty-five stretched A300s. Air Canada followed. Then Federal Express, America West, and TWA. In 1991, Delta placed an order for nine A310s.

Finance had become the final arbiter in the competitive battle with Airbus, with government subsidies providing the advantage in both global and US markets.

The US Commerce Department reported that Airbus Industrie received more than $16 billion in subsidies since its formation. Finally, after six years of talks, US and European negotiators in April 8 1992 announced a tentative agreement that would cap direct European subsidies at 33 percent of new development costs and limit direct benefits - such as new technology developed for the military that could be applied to commercial aircraft - to four to five percent of those outlays. But the subsidy conflict was by no means over.

In the spring of 1982, Airbus announced to the world that it would design a new technology twin, the A320, promising deliveries in 1988.

Boeing had been studying a 150 passenger airliner, but the list of unk-unks - Boeing-speak for unknown-unknowns - was longer than ever, and Boeing paused to digest them. Then, in March 1984 Boeing signed an agreement with Japan to

Below: the Boeing 7J7 only ever appeared as a set of concept drawings.

However, the powerplants that were to drive it did fly on an MD.81 (left) or Boeing 727 (above).

The engine was also known as a propfan, or an open rotor engine,

develop a new competitor to the A320, but no timetable was set and Boeing were keeping their plans close to their corporate chests. Scarcely three months after this agreement was announced, Pan American ordered twenty-eight Airbus airliners, including the new A320, for more than one billion dollars.

Boeing still seemed to be in no hurry to make a decision. At the time, Tex Boullioun made a flat statement, 'The A320 is absolutely impossible economically. It's going to be tough for anybody to come up with a new, small airplane that makes any money.' With such a positive statement from the President of the Commercial Airplane Group, there was a visible relaxation all along the line.

Setting the tone for the airlines to delay their decisions - hopefully to favour future Boeing products - Boeing's Joe Sutter told attendees at the 1983 Paris Air Show that 'We're just sounding a note of caution for people looking at airplanes like the A320, because time marches on and technology marches on. Structures, aerodynamics, systems, and engines are constantly improving.'

It was not until 1985 Boeing announced its market strategy with derivatives of the 737 while taking several strides in technology for a future machine. They visualized technology developments in flight systems, aerodynamics, structural materials, and propulsion that seemed to promise a sixty per cent improvement in efficiency.

The new airliner would be known as the UDF, standing for Unducted Fan, and in Boeing terminology - the 7J7. The design incorporated aft-facing, fuselage-mounted engines, and multiple, counter-rotating blades.

Following its strategy plan, Boeing brought out the 737-300 in 1985. The airliner sold an incredible total of 252 units in the first year. Even so, the question of whether derivatives could stop the A320 challenge remained to be answered. It seemed to be no. In September 1986, Northwest Airlines, long an exclusive Boeing customer, signed an agreement to purchase up to one hundred A320s. That decision was taken in the face of Boeing's next derivative, the 737-400.

As major problems began to mount, Boeing announced a fifteen-month delay for the 7J7. Then, in the fall of 1987, with shock waves reverberating through the Company, the 7J7 went back on the shelf.

It is often claimed that even when Airbus developed the A320 as a competitor to the second-generation 737s, Boeing Management still thought their 737 range was superior to the A320, despite it having a fly-by-wire control system and more computers.

That, however, completely missed out a little-known airliner that only sold in very small numbers, was almost identical to the 737 and was the precursor to the A320.

Marcel Dassault (*b*. Marcel Bloch; 22 January 1892, *d*.17 April 1986) was a French industrialist who spent his career in aircraft manufacturing.

During the mid-1960s, Marcel Dassault, the founder and owner of French aircraft company Dassault Aviation, as well as other organisations such as the French Directorate-General for Civil Aviation (DGAC), examined the civil

Dassault Mercure (above) and A320 (below) both in Air Inter colours.
(Simon Peters collection)

aviation market and noticed that there was no existing aircraft intended to serve short-distance air routes. It appeared there was a prospective market for such an airliner if it were developed. The DGAC was keen to promote a French equivalent to the popular American Boeing 737 and suggested the development of a 140-seat airliner to Dassault.

In 1967, with the backing by the French government, Dassault decided to start work on its short-haul airliner concept. During 1968, initial studies by the company's research team were set around a 110 to 120-seat airliner powered by a pair of rear-mounted Rolls Royce Spey turbofan engines; as time went on, a specification for a 150-seat aircraft with a 1000 kilometre or 540 nautical mile range was developed. The new airliner would attack this market segment at the upper end with a 140-seat jetliner, contrasting against the 100-seat Boeing 737-100 and the 115-seat Boeing 737-200 variants then in production. In April 1969, the development programme was officially launched.

This aircraft was viewed as being a major opportunity for Dassault to demonstrate to the civilian market its knowledge of high-speed aerodynamics and low-speed lift capability that had been developed in the production of a long line of jet fighters, such as the Ouragan, Mystère and Mirage aircraft. The French Government contributed fifty-six per cent of the programme's total development costs, which was intended to be repaid by Dassault via a levy on sales of the airliner, the company also financed the initiative with $10 million of its own money, as well as being mainly responsible for costs related to manufacturing.

Marcel Dassault decided to name the aircraft Mercure - French for Mercury. 'Wanting to give the name of a god of mythology, I found of them only one which had wings with its helmet and ailerons with its feet, from where the Mercure name..' said Marcel Dassault. Stat-of-the-art modern computer tools were used to develop the wing of the Mercure 100. Even though it was larger than the Boeing 737, the Mercure 100 was the faster of the two. In June 1969, a full-scale mockup was presented during the Paris Airshow at Le Bourget Airport. On 4 April 1971, the prototype Mercure 01 rolled out of Dassault's Bordeaux-Merignac plant.

The maiden flight of the prototype, powered by a pair of Pratt & Whitney JT8D-11 turbofan engines, capable of generating up to 6800 kilogrammes of thrust, took place at Mérignac on 28 May. On 7 September 1972, the second prototype, which was powered by a pair of Pratt & Whitney JT8D-15 engines, which would be used on all subsequent Mercures built, flew for the first time. Less than a year later, on 19 July 1973, the first production aircraft conducted its maiden flight. On 12 February 1974, the Mercure received its Type certificate and, on 30 September 1974, was certified for Category IIIA approach all-weather automatic landing, meaning minimum horizontal visibility of five hundred feet and a minimum ceiling of fifty feet. According to *Flight International,* the basic model of the Mercure featured a degree of built-in stretch potential; elements of the design were reportedly capable of supporting the envisioned expanded model with little or no change, including much of the wing, cabin, and the undercarriage, the latter being spaced in order to accommodate the fitting of longer legs to in turn enable larger engines and an elongated fuselage to be later adopted.

Standing side-by-side. Boeing to Airbus

Variant	Pax	Range
First Generation		
737-100	118	1,540 nmi
737-200	130	2,600 nmi
Second Generation		
737-300	140	2,255 nmi
737-400	168	2,060 nmi
737-500	132	2,375 nmi
Third Generation		
737-600	130	3,235 nmi
737-700	140	3,010 nmi
737-800	175	2,935 nmi
737-900	215	2,950 nmi
Fourth Generation		
37 MAX 7	153	3,850 nmi
737MAX 8	178	3,550 nmi
737MAX 9	193	3,550 nmi
737MAX 10	204	3,300 nmi

Variant	Pax	Range
A318	117	3,100 nmi
A319	160	3,750 nmi
A320	190	3,300 nmi
A321	230	3,200 nmi
A319NEO	160	3,750 nmi
A320NEO	195	3,500 nmi
A321NEO	240	4,000 nmi

Dassault tried to attract the interest of major airlines and several regional airlines, touting the Mercure 100 as a replacement for the Douglas DC-9. A few airlines showed some initial interest but only Air Inter, a French domestic airline, placed an order. This lack of interest was due to several factors, including the devaluation of the dollar and the oil crisis of the 1970s, but mainly because of the Mercure's operating range – suitable for domestic European operations but unable to sustain longer routes; at maximum payload, the aircraft's range was only 1,700 km. Consequently, the Mercure 100 achieved no foreign sales. With a total of only ten sales with one of the prototypes refurbished and sold as the 11th Mercure to Air Inter, the airliner represented one of the worst failures of a commercial airliner in terms of aircraft sold.

Later, in order to answer a request from the DGAC, Dassault proposed a Mercure equipped with the new General Electric/Snecma CFM International CFM56; this version came to be known as the Mercure 200. In 1975, contacts were made with Douglas and Lockheed to build and sell the Mercure 200 in the USA, and with SNIAS to build it in France. However, Marcel Dassault was concerned that the CFM56 had not had a single order yet, and production might end before the Mercure 200 could be built. Meanwhile, Douglas proceeded to introduce a stretched version of the DC-9, which was in direct competition for orders with the Mercure 200; accordingly, contacts with Douglas were terminated.

There is more than enough evidence to suggest that lessons learned with the Mercure 100, and design aspects in hand for the -200 series were incorporated into the Joint European Transport (JET) programme set up in June 1997, established by BAe, Aerospatiale, Dornier and Fokker. It was based at the then British Aerospace, formerly the Vickers site in Weybridge, Surrey, UK. Although the members were all of Airbus Industries' partners, they regarded the project as a separate collaboration from Airbus.

This project was considered the forerunner of Airbus A320, encompassing the 130- to 188-seat market, powered by two CFM56s. It would have a cruise speed of Mach 0.84 - faster than the Boeing 737. The programme was later transferred to Airbus, leading up to the creation of the Single-Aisle (SA) studies in 1980, led by the former leader of JET programme, Derek Brown. The study group looked at three different variants, covering the 125- to 180-seat market, called SA1, SA2 and SA3 that was to eventually surface as the A319, A320 and A321, respectively. The single-aisle programme created dissention within Airbus as to design a shorter-range twinjet rather than a longer-range quadjet wanted by the West Germans, particularly Lufthansa. However, works proceeded, and the German carrier would eventually order the twinjet.

In February 1981, the project was re-designated A320, with efforts focused on the former SA2. During the year, Airbus worked with Delta Air Lines on a 150-seat aircraft envisioned and required by the airline. The A320-100 would carry 150 passengers over 2,850 or 1,860 nautical miles - 5,280 or 3,440 kilometres - using fuel from wing fuel tanks only. The -200 had a working centre tank, increasing fuel capacity from 3,429 to 5,154 imp gallons, or 15,590 to 23,430 litres. Airbus settled in a broader cross-section with a 12 feet 2 inches, or 3.7 metres internal width, compared to the 737's 11 feet 4 inches or 3.45 metres. Although more substantial, this allowed the A320 to compete more effectively with the 737. The A320 wing went through several stages of design, finally settling on 111 feet 3 inches, or 33.91 metres.

Boeing had become the world's pre-eminent commercial aeroplane manufacturer in part because it developed a coherent design philosophy that relied on pilots' airmanship as the last line of defence. It made sense in an era when airliners were vulnerable to weather and prone to failures and pilots regularly intervened to keep their machines from crashing. By the 1980s it became apparent that due to engineering improvements, very few accidents were caused by the machines anymore, and almost all resulted from pilot error. This occurred at a time when airlines were being deregulated, and discount carriers were springing up, major new markets were beginning to appear in developing countries, pilots' unions were being busted, pilots' salaries were in steep decline and airmanship globally was being eroded by an increasing reliance on cockpit automation, production-line training and a rote approach to flying.

In the face of these changes, Boeing clung firmly to its pilot-centric designs, but in Toulouse, Airbus was not nearly as shy. Led by an outspoken former military test pilot turned chief engineer named Bernard Ziegler, Airbus decided to take on Boeing by creating what some saw as a robotic new airliner that would address the accelerating decline in airmanship and require minimal piloting skills largely by using digital flight controls to reduce pilot workload, iron out undesirable handling characteristics and build in pilot-proof protections against errors like aerodynamic stalls, excessive banks and spiral dives. Simply put, the idea was that it would no longer be necessary to protect the public from airliners if Airbus could get the airliners to protect themselves from pilots.

The approach was diametrically opposed to Boeing's. Ziegler announced that he was going to build an aircraft that even his concierge could fly. The implicit insult won him the enmity of some French airline pilots, who then as now thought highly of themselves. But his efforts led to the smartest airplane ever built, a

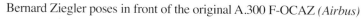

Bernard Ziegler poses in front of the original A.300 F-OCAZ *(Airbus)*

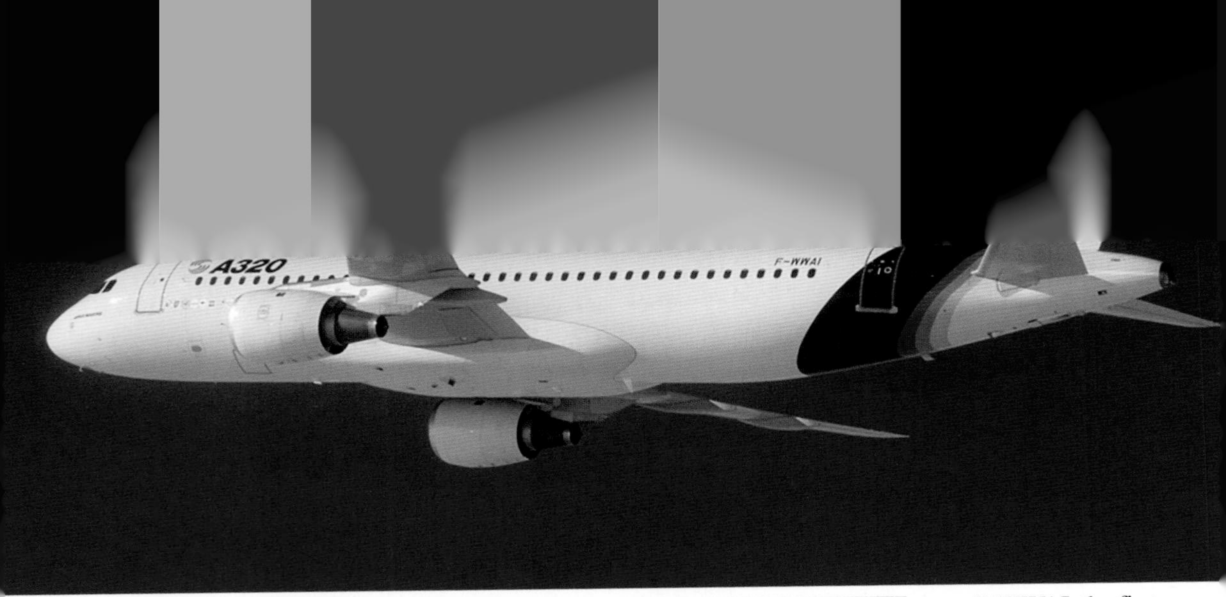

F-WWAI, the first Airbus A320 ever built.

The prototype A320 rolled out on 14 February 1987 and conducted its first flight a week later on 22 February. After 1,200 accumulated hours during 530 test flights, the European Joint Aviation Authorities certified the aircraft one year after rollout on 26 February 1988. Airbus delivered the first A320 to Air France on 26 March 1988. *(both Airbus)*

single-aisle medium-range 'fly-by-wire' masterpiece called the A320 that entered the global market in 1988, led the way to all other Airbus models since and has been locked into a seesaw battle with Boeing's relatively conventional 737s for the past thirty-or-so years.

The 737 and A320 were closely matched: same payloads and performance, same operating costs, the same potential for profit-making.

Then came the Trade War. The European Union and United States have been battling over government subsidies to Airbus and Boeing for at least fifteen years in the world's largest ever corporate trade dispute.

The World Trade Organization found the world's two largest planemakers received billions of dollars of unfair subsidies in a pair of cases dating back to 2004 and is expected to allow both sides to impose tariffs, starting with the US.

The feud that spawned thousands of pages of rulings, triggered threats of tit-for-tat tariffs on goods from airliner parts to whisky and left both sides claiming victory while racking up an estimated $100 million in costs.

Indeed, I covered the feud in detail in my books (see *The Airbus A380 A History* Pen & Sword 2014 and *The Boeing 707 Group,* Pen & Sword 2017) but it is worth re-iterating the timeline here, for the feud had an impact on the reasons why Boeing developed the MAX.

2004 - The USA seeks talks with the European Union and Airbus host nations Britain, France, Germany and Spain over alleged unfair subsidies via government loans. Washington also terminates a 1992 US-European Union agreement covering support for Airbus and Boeing. The EU responds by filing a complaint on US aid for Boeing that includes the many years of US military orders for aircraft whose tooling can be used on civilian versions - such as the KC-135 tanker and the 707 airliner.

2005 - The World Trade Organization launches twin probes into public support for Boeing and Airbus after bilateral negotiations fail.

2006 - Airbus announces a new A350 jetliner for which it will seek further government loans from host nations.

2009 - The World Trade Organization issues an interim ruling that some European aid provided to Airbus violated a ban on export subsidies - a type of aid deemed most harmful and therefore automatically banned.

2010 - The World Trade Organization demands a halt to unfair aid for airliners including the Airbus A380. It says some government loans for the jet amount to prohibited export subsidies. At the same time it rejects a US request to include aid for the newer A350 in the case.

2011 - The European Union loses an appeal and is given until December to comply. However the World Trade Organization drops its finding that the A380 loans are in the prohibited category, so softening its earlier ruling.

Both companies clash over the scope of the World Trade Organization finding. Boeing says the World Trade Organisation faulted $18 billion of subsidies to Airbus including $15 billion in loans. Airbus says the amount of subsidy embedded in the loans is far smaller, but neither side can publicly back up its claims since details are redacted.

In the European Union's counter-case alleging $19 billion of support for Boeing from the US government, NASA and various states and municipalities, a separate World Trade Organisation panel partially backs the EU and rules against aid for Boeing worth at least $5.3 billion.

2012 - World Trade Organisation appeal judges broadly uphold the ruling against US support for Boeing. Both sides say they have complied with the World Trade Organization's rulings while accusing the other side of failing to do so. This latest disagreement kicks off a new compliance phase in the dispute, including appeals, that lasts another six years.

2013 - Boeing announces the twin-engined 777X and agrees to build it in Washington state shortly after the local legislature agrees $8.7 billion in new tax breaks.

2014 - The European Union opens a second front in the trade battle by launching a separate complaint against the 777X tax breaks granted by Washington state and this time chooses a faster, all-or-nothing approach by targeting them purely as prohibited subsidies - without the usual fallback of a second, softer claim.

2016 - After a year-long lull in the main dispute, the World Trade Organization says the European Union failed to comply with its earlier rulings on Airbus. It also agrees for the first time to target aid for the new A350 but rejects US calls to put this in the prohibited category.

In November, the World Trade Organisation rules tax breaks surrounding the development of the Boeing 777X - the subject of the EU's second case - did fall into the more severe prohibited category.

2017 - World Trade Organization appeal judges reverse the ruling that the 777X tax breaks are in the prohibited column, bringing an abrupt halt to the European Union's second case after only two and a half years - a quick turnover by the standards of the main legal battle. The European Union's original case against aid for earlier Boeing projects - including an earlier version of the same tax breaks - continues.

In that main European Union case, the World Trade Organization largely clears the United States of maintaining unfair support for Boeing but says it has failed to withdraw the earlier Washington State tax breaks.

The European Union appeals this decision, but the World Trade Organisation does not change its stance in a follow-on ruling, published the same year.

2018 - In May, the Word Trade Organization again rules that the European Union has failed to halt all subsidies to Airbus and that these continue to harm Boeing. The United States threatens sanctions on billions of dollars of European products. Both sides enter arbitration to determine the scope of tariffs.

2019 - In March, the World Trade Organization says the US has again failed to halt subsidized tax breaks to Boeing in Washington state. Once again, the two sides disagree widely in public over the amount of subsidy faulted by the World Trade Organisation.

In April, the USA issues a list of $21 billion worth of European Union products from which any World Trade Organisation-approved tariffs could be drawn, ranging from aircraft to food and handbags, so as to counteract $11.2 billion of harm it says European Union subsidies cause the US each year.

The European Union issues its own $20 billion list of US imports that could face tariffs for damage from US subsidies.

In June, US sources said they are open to negotiations on an enforceable mechanism that could allow Airbus to receive some government funding on commercial terms while addressing the issue of Washington State tax breaks.Both sides accuse the other of refusing to negotiate any settlement.

In July, the US adds $4 billion of items to the basket of products from which any tariffs against the European Union could be drawn.

So even at the start of the trade war, it was clear that Boeing was prepared to use every weapon in its arsenal to counter any commercial threat from Airbus.

Chapter Five

The -300/700 Series

By the early1980s, it became apparent that the 737 was in need of a serious update to make use of new technologies to meet airline requirements better and also to satisfy growing environmental concerns regarding noise. Boeing designers began working on a 'new generation' 737 and the go-ahead for the 737-300 was authorised on 15 March 1981.

The 737-200 was succeeded in 1984 by the 737-300. This was a much quieter, larger and more economical aircraft and contained a host of new features and improvements. The new model featured many aerodynamic, structural, cockpit and cabin features developed for the new generation 757/767.

That said, one of the key objectives was to have a high degree of commonality with the 737-200; the achieved figure was 67% by part count. This gives saving for airlines in maintenance, spares, tools for existing 737-200 operators. Also, the aircraft was designed to have similar flying qualities, cockpit arrangements and procedures to minimise training differences and permit a common type rating.

Dimensionally, the wingspan for the new generation was increased slightly, and the leading edge slats were extended from the engines outboard to the wingtips. The wing had to be strengthened to carry new engines, and so the increase in high-lift devices helped to compensate for the increase in weight and keep the approach speed down. The 737-300 was longer, permitting an average two-class capacity of 128 or an all-economy layout of up to 149 seats.

The sole powerplant was the CFM-56, the core of which is produced by General Electric and was virtually identical to the F101 as used in the Rockwell B-1. SNECMA produced the fan, IP compressor, LP turbine, thrust reversers and all external accessories.

The main problem was the size of the engine for ground clearance; this was overcome by mounting the accessories on the lower sides to flatten the nacelle bottom and intake lip to give the 'hamster pouch'

BOEING 737-300

Length:	109ft 7 in (28.88 metres)	**Fuel (max):**	6,295 US gals (23,829 lit)
Height:	36 ft 6 in (11.12 metres)	**Range:**	2784 miles (4481 km)
Span:	94 ft 9 in (28.88 metres)	**Cruise Speed:**	0.745 mach
Empty Weight:	76,490 lbs (32,881 kg)	**Pax Capacity:**	128-149
Max Weight:	139,500 lbs (63,276 kg)	**Powerplants:**	2 x CFM56-3C-1

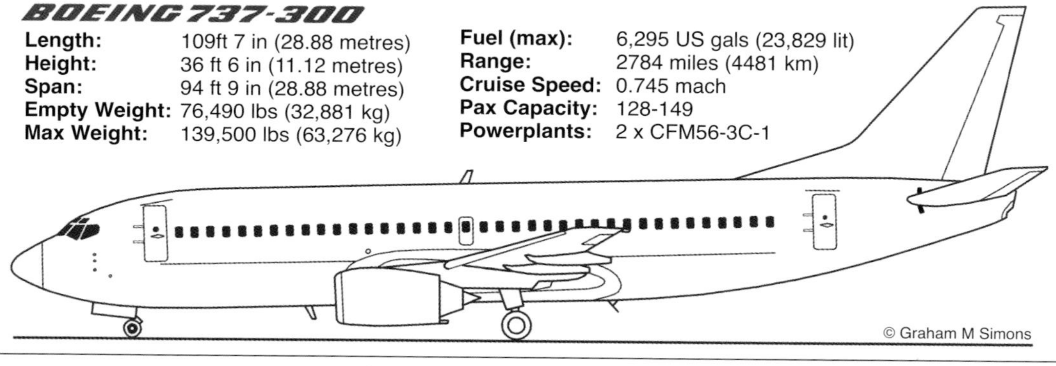

© Graham M Simons

A promotional image of the series 300 sales demonstrator at the 1984 Paris Air Show.

look. The engines were moved forward and raised level with the upper surface of the wing and tilted five degrees up. This not only helped with ground clearance, but also directed the exhaust downwards, reducing the pylon heating and giving slight vectored thrust to assist take-off performance. The CFM56-3 proved to be almost twenty percent more efficient than the JT8D.

Two sections were added to the basic -200 fuselage; a forty-four-inch section forward of the wing and a sixty-inch section aft of the wing. Composite materials were used on all flight controls to reduce weight. Aluminium alloys were used in areas such as wing spars, keel beams and main landing gear beams to improve their strength by up to twelve per cent so increasing service life.

The wings were extensively redesigned to enhance low-speed performance and cruise efficiency. The chord of the leading edge outboard of the engines was extended by four point four per cent; this reduced the wing upper surface camber forward of the front spar to increase the critical Mach Number, thereby giving better transonic airflow characteristics and improved buffet margins. The span was increased by a wingtip extension. These changes impacted on high-speed performance and turbulent air penetration speed was increased.

High lift characteristics were also enhanced by re-sequencing the slats and flaps. The leading edge slat radius was also increased which gave a two-and-a-half knot reduction in Landing Reference Speed over a -200 at the same weight. Other changes to the wing structure included strengthened materials and corrosion protection.

These were by no means the only changes. On the flight deck, there was now a Flight Management System with fully integrated Digital-Flight Control System, Autothrottle, Flight Management Computer, Dual laser gyro Inertial Reference System and Electronic Flight Instrument System (EFIS) using cathode-ray tube - TV - (CRT) displays.

• Performance: Higher ceiling at 37,000 feet; higher Maximum Take-Off Weight; twenty per cent lower fuel burn than the series 200.

• Engines: High bypass (5:1), 18,500-23,500 pounds, CFM56-3.

• Fuselage: forty-four inch forward body extension, sixty-inch aft body extension, strengthened body skins and stringers.

• Wings: Span increased by eleven-inch wingtip extensions; Modified slat aerofoil, flipper flaps and flap track fairings. Increased fuel capacity to 35,700 pounds

• Tail: Dorsal fairing added for stability during asymmetric conditions; Stabiliser extended thirty inches due to increased fuselage length.

• Flight Controls: New slats to reduce approach speeds to within a few knots of the much lighter -200. Additional ground spoiler. Stabiliser tip extension.

• Nose Gear: Lengthened six inches and repositioned to help provide the same

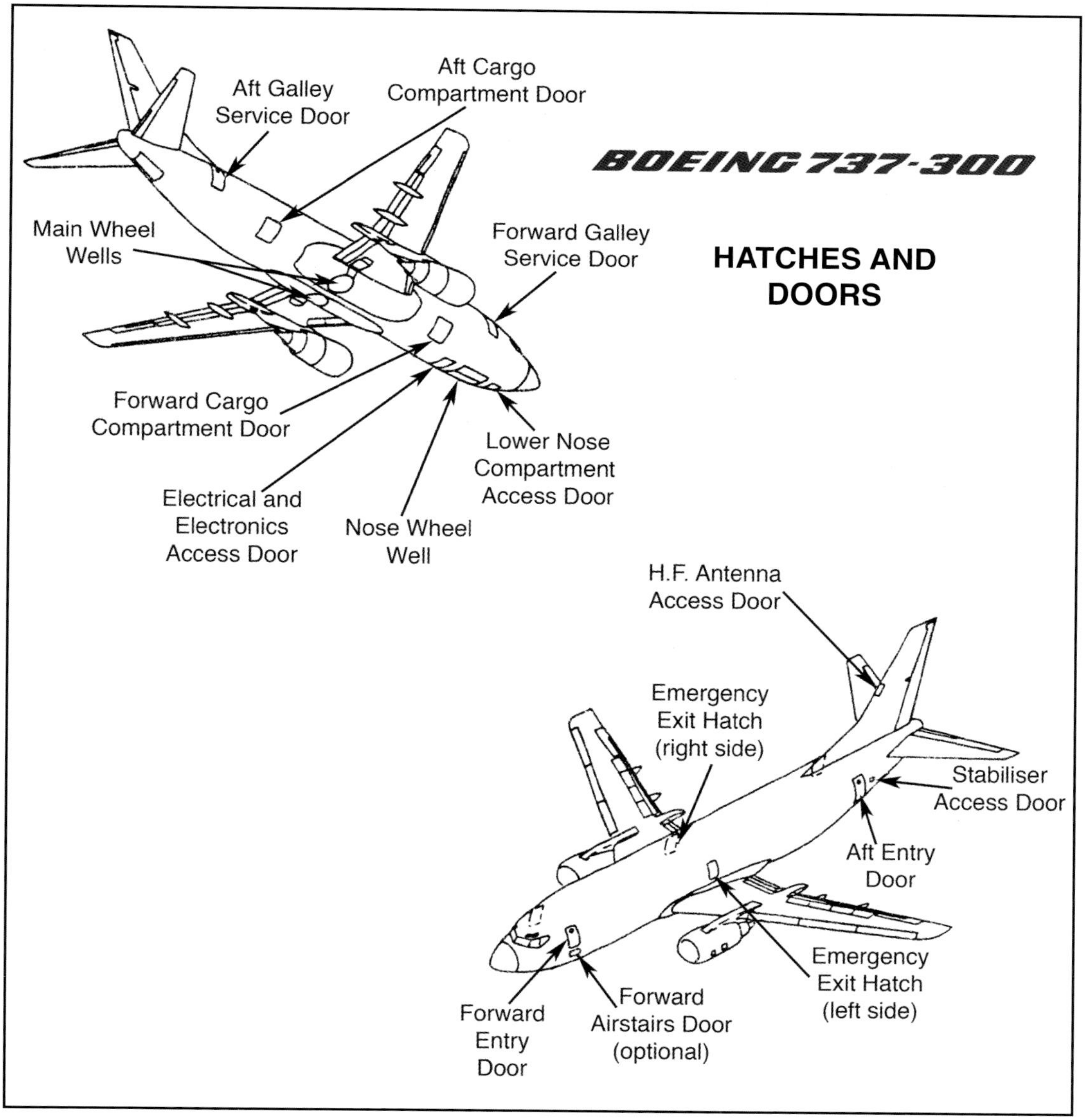

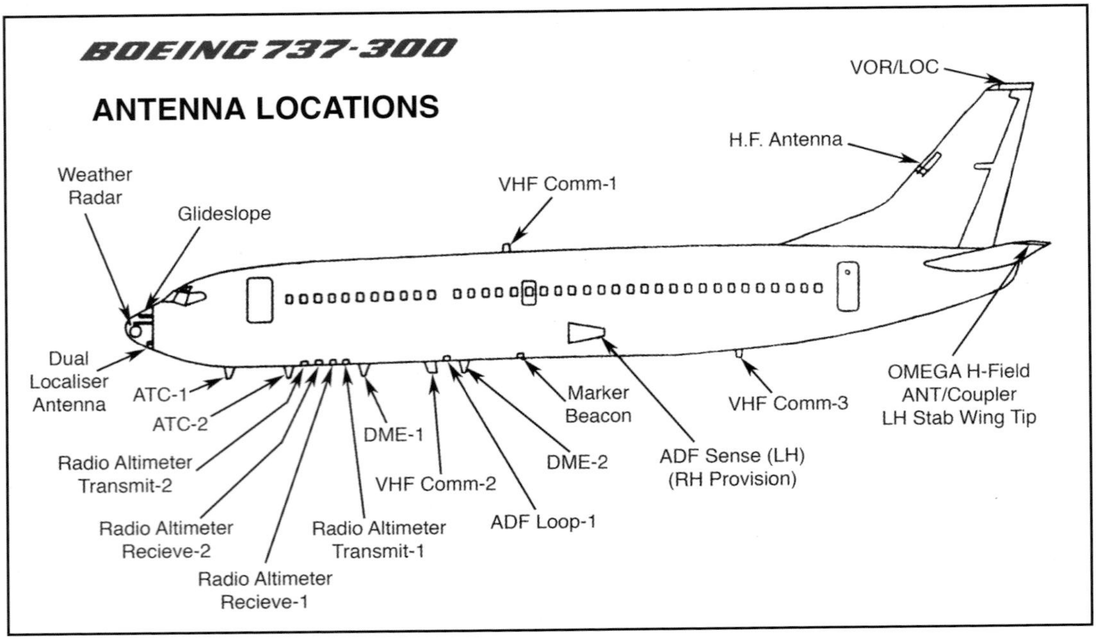

engine inlet ground clearance as the -200.

- Main Gear: Strengthened wheels, forty or forty-two-inch tyres and better brakes for the increased Maximum Take-Off Weight.
- Flightdeck: EFIS CRT displays replacing conventional instruments.

Various changes and customer options became available on the Classics over their production run, that included:

- Taxi-light extinguishing automatically with gear retraction.
- Addition of a strobes AUTO position.
- Light test not illuminating engine,

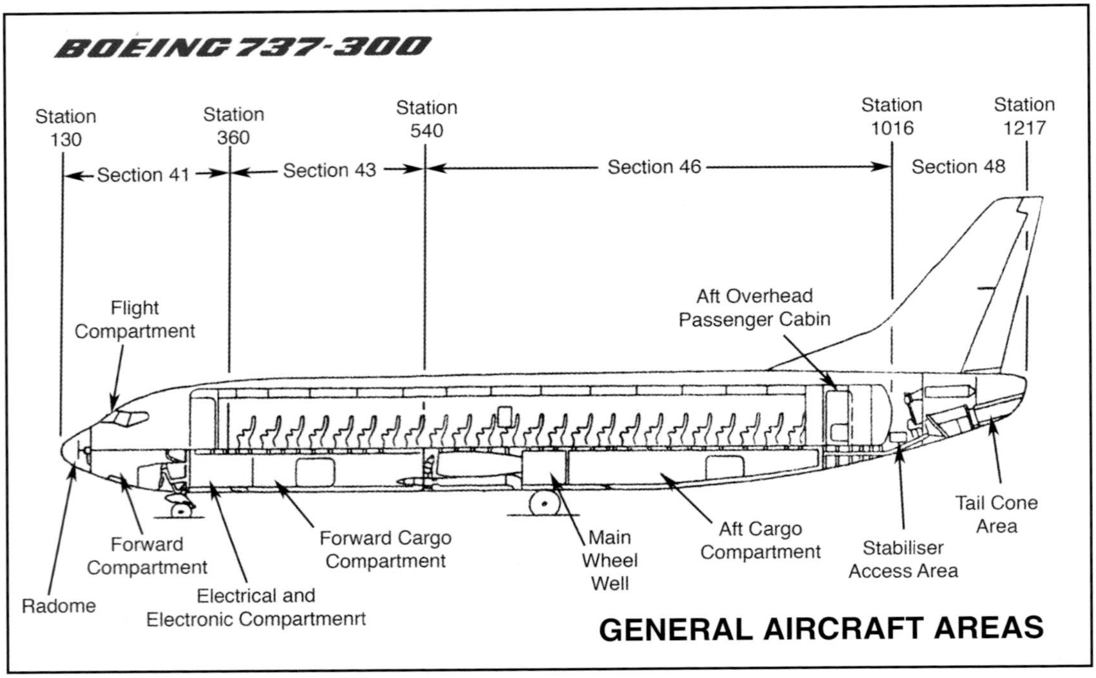

Above: Dan-Air's 737-3Q8 G-SCUH was leased from International Lease and Finance Corporation. *(Dan Air Staff Association)*

Left: A detailled view of Dan's airstairs as fitted to their 737-300s. The escape slide is in the bustle attached to the door, and the stairs themselves and their mountings come out of an underfloor hatch that is re-enforced around the fuselage skin. *(author)*

Auxilliary Power Unit (APU) and wheel-well fire warnings.

• Overheat/Fire system test will only show a fault, not determining whether open or closed circuit.

• Cabin pressure control system (CPCS) and indicators replaced by digital system (DCPCS) which had no standby mode but has a second (alternate) automatic mode.

• A Sundstrand APU with no EGT limits and shorter restart intervals.

• Extra automatic DC fuel pump for APU starts.

• Various Comms options such as VHF3, HF, SELCAL, ACARS & TCAS.

• Option of VSCF instead of CSD.

• Dripsticks replaced by floatsticks.

• Option of a fourth fuel tank.

• Refuel preset facility on refuelling panel.

• Alternate nose wheel steering (from hydraulic system B).

• Side window heat now also heats window 3.

• Third automatically tuned DME radio for FMC position.

• Round dial engine instruments replaced by EIS panels.

• APU battery.

• Aspirated TAT probe.

Boeing resurrected the N73700 registration for the 737-300 prototype (c/n 22950), the 1,001st 737, rolled out at Renton on 17 January 1984. With Boeing test pilot Jim McRoberts in command, the aircraft made a successful first flight on 24 February. Immediately afterwards, the prototype and the first two production -300s were involved

The flightdeck of easyJet 737-33V HB-IIT *(author)*

in a test and certification programme before the type could be introduced by launch customers Southwest Airlines and USAir.

The flight test programme lasted nine months, and the three aircraft flew a total of 1,294 hours. There were some problems with engine/wing flutter, but these were overcome with the addition of a 100-pound mass balance at each wingtip.

Initial orders were slow, but within two years, the -300 proved to be a huge success and orders flooded in. By the time production of the -300 ceased in late 1999, some 1,113 had been delivered - just one less than the -200 series.

The company saw that the constantly changing needs of the airlines meant it had to react to these requirements in the face of increasing competition from its European rival, Airbus Industrie. What Boeing needed was an aircraft with a capacity between that of the 737-300 and its model 757. The result was the stretched 737-400, ten feet longer than the -300 and able to seat up to 170 passengers in a single-class layout.

Next Generation: 737-300X

The Boeing 737-X programme was launched on 29 June 1993, with a 63 aircraft order from Southwest Airlines for the 737-300X. This became the 737-700, nine inches longer than the original 737-300, seating up to 149. The main differences created with the 737 Next Generation (NG) were:

- Performance: Faster cruise 0.78 M, Higher ceiling 41,000 feet, Lower take-off and approach speeds, Higher Maximum Take-Off Weight and lower fuel burn.
- Engines: Full Authority Digital Engine Control (FADEC) controlled CFM56-7 with two-and-a-half degree nozzle tilt, redesigned struts, improved nacelles with increased airflow and improved noise treatment, 7% more fuel efficient than CFM56-3.
- Fuselage: Strengthened for increased tail loads and design weights, new wing-body strake.
- Wings: New airfoil section, 25% increase in area, with a semi-span increase, chord increase, raked wing-tip, larger inspar wing box with machined ribs.
- Fuel Tanks: Main tanks smaller at 8600 pounds each but centre tank much larger giving fuel capacity of 45,800 pounds.
- Tail: Four feet eight inches taller, sixty square foot root insert, modified rudder, segmented rudder seals, digital yaw

Continental Airlines' 737-3TO is seen at Cleveland. *(author)*

'go' - and yes, that is spelt correctly in pure lower-case letters as was the vogue of the time - was British Airways' short lived attempt to counter the growing influence of the low-cost carriers that were starting to appear across Europe in the 1990s. Here G-IGOL is seen on approach to London Stansted. Go Fly, styled and trading as go was formed in May 1998 with Barbara 'Babs' Cassini as its head. The airline was eventually sold to easyJet in May 2002. *(author)*

damper.

- Flight Controls: Increased elevator Power Control Unit capability, aileron and tab span increase, new double slotted continuous span flaps, new leading-edge Krueger flaps, additional slat, additional spoiler.
- Nose Gear: Stroke increased 3.5 inches to relieve higher dynamic loads and wheel well extended three inches forward.
- Main Gear: Longer to reduce tailstrike risk, one-piece titanium gear beam, 43.5 inch tyres, digital antiskid.
- Flightdeck: Six programmable Liquid Crystal Displays, replacing EFIS CRT displays and most conventional instruments.
- Systems: Most systems developed particularly: electrics, powerplant and navigation.

The NGs had 33% fewer parts than the Classics, which reduced build time. One of the main production differences with the NG was the single moving assembly line. This had the capacity to produce twenty-one aircraft a month with a flow time of just thirteen days.

Boeing introduced every innovation possible while still staying within the

British Midland's Series 33A G-OBMJ on approach to East Midlands. *(author)*

'derivative' rules. Widening of the fuselage or introducing fly-by-wire would have forced the design into the 'new' airliner category, with attendant costs of hundreds of millions of dollars for recertification. Further, commonality - certainly the greatest attribute of the 737 - would have been highly compromised.

Gordon Bethune, vice president and general manager of the Renton Division of the Commercial Airplane Group, said, 'Our customers were quite clear about what they wanted - and didn't want - in the 737-X. They asked us to change as little as possible unless it related to range, noise, speed, or seating capacity.

737 Fire Liner

Specialist firefighting company Coulson Aviation of Canada converted six ex-Southwest 737-300s into aerial firefighters, known as 'Fire Liners'. The air tanker was intended as a multi-use aircraft with the ability to carry passengers. Britton 'Britt' Coulson, vice-president aviation with Coulson Aviation, said: 'It's the only air tanker in the world that can carry its load of retardant at high altitudes and high speeds. It takes advantage of the airframe. And it's one of the few air tanker modifications in the world that doesn't limit the airplane at all. This is the largest ever modification of a Boeing 737. The Fireliner represents the newest generation of air tankers, not only fast but also versatile, capable of transporting seventy-two firefighters in between its twin tanks, each of which has a capacity of more than 4,000 US gallons'.

'With a full retardant load and 4.5 hours of fuel we are so far under max gross weight we are going to leave the full interior and galleys in even when just in airtanker mode.'

Conversion of the first aircraft, coded T-137, began in 2017 and took 43,000 man-hours. It was then repainted at Spokane. The gravity-based tanks of the retardant system were complete in December 2017. The first conversion was

The former Southwest Airlines 737-3H4W N617SW, now registered as N137CG for Coulson Aviation but with New South Wales Rural Fire Service titles is seen on approach. The distinctive colour scheme at least partially conceals the underbelly staining caused by the red-coloured fire retardent slurry. The fluid is coloured that way so that pilots can see where they have previously dropped. One of the most commonly used brands of retardant, Phos-Chek, uses red iron oxide to dye its product, a mix of ammonia and nitrates that can keep trees and other flammable materials from combusting. The aircraft also carries T-137, its Coulson tanker number under the forward fuselage.
(Liam Cannuck collection)

done by Coulson Aircrane Canada. The second aircraft, Tanker 138, arrived in April 2018 at the Port Alberni Facility and was converted with the RADS XXL tanking system for aerial firefighting.

On 13 July 2018 air tanker T-137 made its first drops while flying out of San Bernardino, California. Britt Coulson said, 'The 4,000 USG RADS-XXL/2 performed perfectly, as did the airplane. Our flight crew couldn't have been happier with the handling characteristics, and our split tank worked as designed with no Centre of Gravity shift during the drop.' Once flight testing with the FAA and static tests of the tank system was complete Coulson worked with the Forest Service to schedule the grid test to assess the characteristics of the retardent drop.

The Fire Liner was deployed to fight bushfires for the first time on 22 and 23 November 2018, in the port city of Newcastle, north of Sydney, Australia. The modified airliner dropped several loads of fire retardant on the out-of-control blaze in the affected area, assisting firefighters on the ground.

Nicknamed 'Gaia' - but also named 'Marie Bashir' as a tribute to Dame Marie Roslyn Bashir, AD, CVO, the former and second longest-serving Governor of New South Wales - the aircraft, which was valued at around $7 million, put under contract with the New South Wales Rural Fire Service (RFSA), the state's volunteer-based firefighting agency.

The type is known as 'Large Air Tanker' in Australia.

CATBird

The Lockheed Martin CATBird was the former Lufthansa Boeing 737-330 D-ABXH, highly modified into an avionics flight testbed aircraft and flying under the registration N35LX. The name is an adaptive acronym, from Cooperative Avionics Test Bed; coincidentally, CATBIRD was Lockheed's ICAO-designated company callsign. The aircraft was modified in order to provide an economical means of developing and flight testing the avionics suite for the Lockheed Martin F-35 Lightning II.

CATBird had a distinctive appearance; so much so that with all the aerodynamic changes made to the airframe it required a supplementary type certificate before it could extend its flight envelope. These modifications included the addition of a nose extension to simulate that of the F-35, a forty-two-foot long spine on the top of the fuselage, a three-metre canoe-shaped pod on the bottom to accommodate electronics and twin 3.6-metre sensor wings forward of the main wings so as to replicate the leading edge of the F-35's wings.

The aircraft interior included an F-35 cockpit installed with sensor inputs to be displayed as they would be in the fighter itself. The rest of the interior housed equipment racks for the avionics equipment, and twenty workstations for technicians to assess the performance of the avionics.

The aircraft was modified under contract by BAE Systems Inc. at their facility at the Mojave Spaceport. Work began in December 2003, and the aircraft began post-modification taxi tests in November 2006. The first flight took place on 23 January 2007 at Mojave. After the initial flight test programme conducted at Mojave, on 2 March, the aircraft was ferried to Lockheed's Fort Worth facility for Phase 2 of the modification programme, which installed the flight test stations and actual avionics and sensor systems to be tested.

In 2014 CATBird software test station was upgraded by Northrop Grumman with Tech Refresh 2 hardware which gave the CATBird capability to test F-35 Block 3 software.

Showing just how strange some of the 'lumps and bumps' modifications to the 737 test-beds could make the airliner look are these two views of the Lockheed Martin CATBird. *(Kirk Smeeton Collection)*

737-700

First flown on 9 February 1997, this was the first of the NGs to take to the skies, and was the equivalent of the 737-300. Just as with the earlier designs, there were a number of sub-variants.

737-700C (Convertible)

Fitted with a 3.4 x 2.1m side cargo door, it could carry 41,420 pounds of cargo on eight pallets. The ceiling, sidewalls and overhead bins remained in the interior while the aircraft was configured for cargo. It had the strengthened wings of the 737-800 to allow higher zero fuel weights.

737-700QC (Quick Change)

A -700C with pallet-mounted seats. This

BOEING 737-700

Length:	110 ft 4 in (33.63 metres)	**Fuel (max):**	6,875 US gals (26,024 lit)
Height:	41 ft 2 in (12.55 metres)	**Range:**	3751 miles (6037 km)
Span:	112 ft 7 in (33.31 metres)	**Cruise Speed:**	0.785 mach
Empty Weight:	84,100 lbs (38,147 kg)	**Pax Capacity:**	128-149
Max Weight:	154,500 lbs (70,080 kg)	**Powerplants:**	2 x CFM56-7B

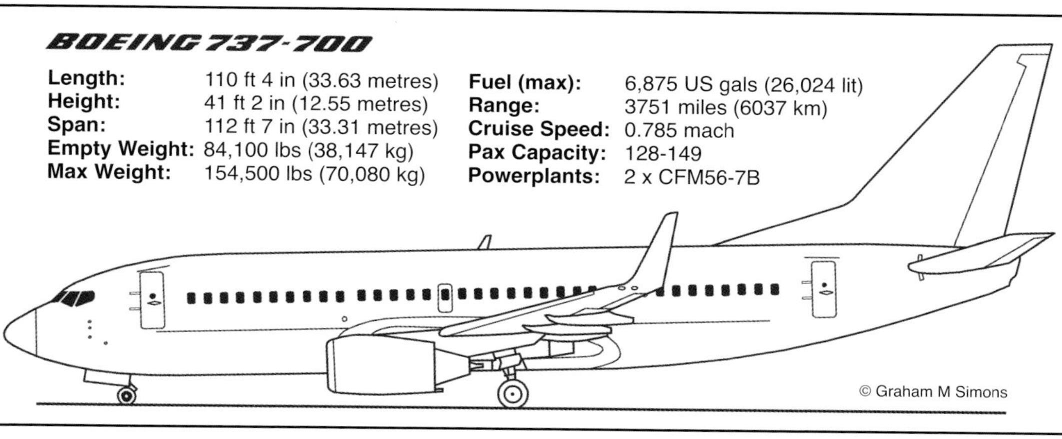

© Graham M Simons

reduced the conversion time from passenger to freighter configuration, and vice-versa, to less than an hour.

737-BBJ

A corporate version of the 737-700 dubbed the Boeing Business Jet (BBJ) was launched on 2 July 1996 as a joint venture between Boeing and General Electric. It combined the fuselage of a 737-700 with the strengthened wings and undercarriage of the 737-800 to carry more fuel - up to twelve fuel tanks, giving 83,000 pounds of fuel could be fitted as a customer option.

The first aircraft, N737BZ, was rolled out at Renton on 26 July 1998, and took to the air for the first time on 4 September. FAA and JAA certification was rapid, coming on 29 October, and the concept has been a resounding sales success. Despite the availability of many smaller business jets, none can offer the accommodation found aboard the BBJ, thanks to its much larger cabin. For example, the BBJ could be equipped with a conference room so the aircraft offers large corporations, the facility to do business on the move.

Externally BBJs differed from standard production 7/8/900s by having various windows blanked to accommodate interior fittings and more antennas for comms equipment; eventually, all had winglets for range. All versions of BBJ only required one overwing exit each side because of their low maximum seating capacity.

The BBJ2 has the 737-800 fuselage, wings and undercarriage. It has twenty-five per cent more cabin space and twice the cargo space or aux fuel tank space of the BBJ 1. The BBJ3 is based on the 737-

Photographs of the interiors of the Business Jets are notoriously hard to come by. This is usually because the operators do not want the shareholders or customers of their clients aware of the ostentatious luxury that their executives are travelling in, as this artists impression shows!

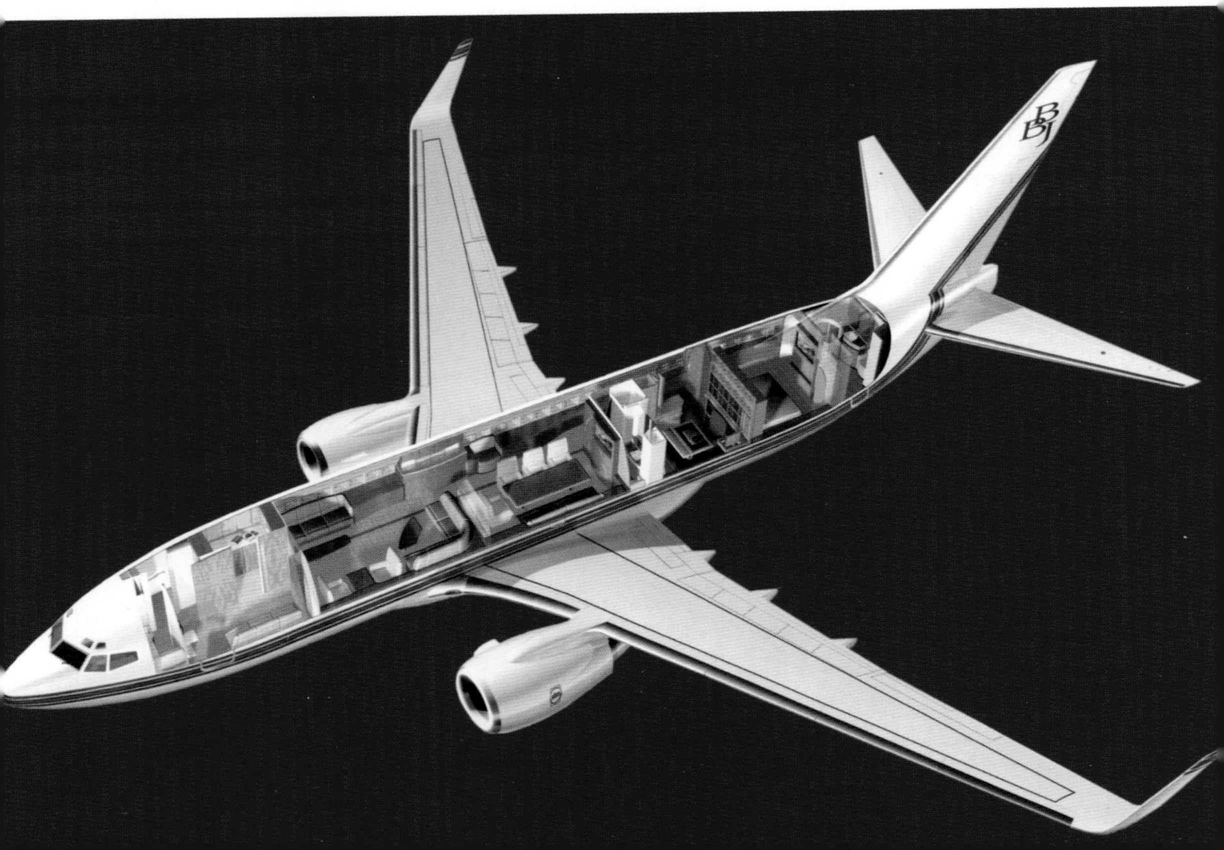

900ER and was available from mid-2008. It has 1120 square feet of cabin space and a range of over 5400 nautical miles with five auxiliary fuel tanks. Boeing also offered a convertible cargo version of the BBJ, called the BBJ C, based on the 737-700

The cockpit was different from that on other 737NGs and was nearer to that found on the 777. Among the more obvious features were the Rockwell Collins Flight Dynamics HGS 4000 Head-Up Display (HUD) System, designed to help pilots during low visibility. Other standard flight deck features included dual global positioning system (GPS), dual HF radios, triple VHF radios, two Smiths Industries flight management computers, six Honeywell liquid crystal display screens, an enhanced ground proximity warning system and a TCAS 2 traffic collision avoidance system.

After roll-out and flight testing at Renton and Boeing Field was completed, the 'green' airframes were delivered to PATS Aircraft Systems, LLC in Georgetown, Delaware. They are a company which designs, manufactures, and certifies aerospace components and systems. The Company offers auxiliary fuel systems, avionics support structures, avionics racks and trays, auxiliary power unit installation kits, aircraft air-stair systems, torque limiting box extractors, and electrical interface systems. At PATS the 737 BBJs were fitted with auxiliary fuel tanks before being flown on yet again to the customer's desired completion centre for painting and fitting out the interior.

The first customer was Boeing partner in the project, General Electric who placed an order for two machines in July 1996, the first delivered on 23 November 1998. By November 2002, some 78 BBJs had been ordered, of which 69 had been delivered. The BBJ featured two CFM56-7 turbofans rated at 26,400 pounds of thrust and the standard -700 fuel capacity of 5,725 gallons. However, up to nine extra fuel tanks could be fitted in the cargo hold space, boosting capacity to a maximum of 8,905 gallons.

One BBJ - N737ER - had been designed for medical evacuations and charter operations and flew a record 6,854 nautical miles from Seattle to Jeddah, Saudi Arabia in fourteen hours twelve minutes. The aircraft still landed with 6,000 pounds of fuel remaining!

A distinctive feature of the BBJ was the blended winglets, developed by Dr L B Gratzer of Aviation Partners, which was

All Nippon Airways' 737-781ER(W) JA13AN seen taxiing out for another corporate flight. *(ANA Business Jet Co. Ltd.)*

Two views of 737-7P3 BBJ VQ-BTA at the 1500 metre short runway at St. Gallen Altenrhein Airport in 2017. The dusk-take off although hard to shoot, gives the whole thing atmosphere. The aircraft was thought to have been operated by Gulf Wings. *(Kirk Smeeton Collection)*

introduced in 2000. They offered improved cruise efficiency and improvements to take-off and climb performance, thus saving fuel. They became a standard feature of the BBJ family and were also optionally available on the -700 and -800 series 737NGs. Winglets in various forms became a standard fitment.

Various combinations of interior were offered and were invariably most luxurious, including combinations of lounges, bedrooms, showers, galley, gym/exercise room, private suite, conference room or up to twenty-four first-class sleeper seats. With the cabin considerably wider than other traditional business jets, the BBJ was able to offer far more flexibility in internal layout.

Considerable interest continued to be shown in the BBJ, although sales of the larger BBJ2, announced in October 1999 and based entirely on the -800 airframe, was somewhat limited. The BBJ2 offered greater range, twenty-five per cent more cabin space and double the hold space.

737-700ER (Extended Range)

This version is essentially an airline BBJ with the stronger -800 wings and higher MTOW. It has up to nine auxiliary fuel tanks, giving a total capacity of 10,707 US gallons and a range of 5,510 nautical miles. It entered service in 2007.

This led to something of an abnormality. Since the -700ER was based on the Boeing Business Jet - itself a NG version of the 737-200 Executive and BBJ's were never certified for production winglets it was not

A good example of a white tail scheme and its use is this shot of another white tail airliner 737-33V F-GZTU of ASL Airlines, caught by the camera climbing out of Corfu International Airport 'Ioannis Kapodistrias' in 2018. *(author)*

ASL Airlines, a subsidiary of ASL Aviation Holdings, is a French airline, certified IOSA. For 30 years, it has a dual activity of passenger transportation and cargo transportation. It is the heir to L'AÉROPOSTALE and it has built its reputation on quality of service in favour of its clients: carriers, postal and express freight operators and tour operators. The full, complicated colour scheme is shown on this publicity picture of 737-33A F-GIXB. *(ASL Airlines)*

surprising that winglets - expected on a long-range aircraft - could not be fitted to the 737-700ER. This, in turn, led to the somewhat strange situation where the first task after the aircraft was built and test flown was to retrofit it with aftermarket winglets before delivery! This bizarre situation was finally resolved in 2008 when the BBJ - and therefore the 737-700ER - was certified for production rather than retrofit winglets.

Not wishing to be left out, the military also went for NG 737s. First to fly on 14 April 2000 the first of a number of what became known as C-40 'Clippers' took to the skies. All have the -700 fuselage combined with the stronger -800 wing and landing gear, similar to a BBJ1.

Soon after the announcement of the series 700, Southwest Airlines immediately showed interest and ordered sixty-three -700s with options to double that amount. Boeing authorised the production go-ahead for the -700 on 17 November 1993.

The FAA awarded type certification for the -700 on 7 November, 1997; the -800 was next on 13 March, 1998, and the -600 on 18 August, 1998. The larger -900, N737X, made its maiden flight on 3 August 2000, and received its FAA certification on 17 April 2001.

Boeing also offered airlines a convertible cargo/passenger version, and a quick-change option using the palletised seat concept tried with the -200 series. The first 737-700C was delivered on 2 November 2001, and featured the strengthened wings used on the Boeing Business Jet, a revised cargo door and a new cargo handling system. The 737-700C was also snapped up by the US military, and the US Navy has adopted the type as the C-40A to replace its fleet of Douglas C-9 Nightingales.

The success of the NG series surprised Boeing, which was receiving orders at a steady and unprecedented rate. In many cases, existing 737 operators were replacing or supplementing older 737 models, but new customers continued to surface. A visit to Renton in late 2000 revealed 737s rolling off the production line at the rate of one a day, with up to sixteen aircraft on the production line at any one time. Production rates continued at this pace until 11 September 2001, terrorist attacks on Washington and New York but the subsequent downturn in commercial aviation in the aftermath caused Boeing to scale back production and lay off many workers.

One event occurred on 17 April 2018 that had implications for the entire NG fleet. A Southwest Airlines 737-700, N772SW, experienced an engine failure while climbing out of New York La Guardia at 10:43 a.m. about twenty minutes after takeoff, as the aircraft was passing through 32,500 feet. The flight was bound for Love Field, Dallas, Texas. A fan blade was thrown from the No.1 engine into the fuselage smashing a passenger window at row 14, killing a passenger who was partially sucked out of the aperture. A further eight passengers suffered minor injuries. As a result of the engine failure, the flight crew conducted an emergency descent and diverted to Philadelphia International Airport

The first NTSB briefing described how multiple aural alerts and warnings sounded on the flight deck. The two pilots donned oxygen masks and reported to air traffic control that they had a No.1 engine fire, were operating on a single engine and were initiating an emergency descent. 'Because they were concerned with potential aircraft controllability issues, they elected to land the airplane with flaps 5 instead of the normal flap setting for a Boeing 737, which would be either flaps 30 or flaps 40'.

The engine fan blades had accumulated over 32,000 engine cycles since new and 10,712 cycles since the last overhaul. NTSB Chairman Robert Sumwalt said in a statement on 19 November 2019: 'The accident demonstrates that a fan blade can

TUIfly 737-7K5(W) D-AHXE in what is known as white tail colours comes in for a landing. From the 1970s, the idea of a plain overall colour idea began to spread, largely in the form of liveries in which white was the dominant colour. A side benefit was that it helped airline asset management. It did so by facilitating the hiring-out of individual fleet members during seasonal traffic peaks or troughs. Overall white aircraft could readily accept major elements of lessee liveries such as logos and brand name stickers, and could equally rapidly revert to lessor liveries on return. *(author)*

fail and release differently than that observed during engine certification testing and accounted for in airframe structural analyses. It is important to go beyond a routine examination of fan blades; the structural integrity of the engine nacelle components for various airframe and engine combinations need to be ensured'.

The NTSB said the engine failure was caused by a broken fan blade, and the board went on to order a series of seven recommendations, three of which had inplications for Boeing:

• Require Boeing to determine the critical fan blade impact location(s) on the CFM56-7B engine fan case and redesign the fan cowl structure on all Boeing 737 next-generation-series airplanes to ensure the structural integrity of the fan cowl after a fan-blade-out event.

• Once the actions requested in Safety Recommendation are completed, require

Two views from the NTSB of the damage caused by the catastrophic failure of the fan disc of the no.1 engine (CFM56-7B24) which suffered an uncontained failure. *(both NTSB)*

Transavia's 737-7K2(W) PH-XRE comes in to land at Xios, Greece. *(author)*

Boeing to install the redesigned fan cowl structure on new-production 737 next-generation-series airplanes.

• Once the actions requested in Safety Recommendation are completed, require operators of Boeing 737 next-generation-series airplanes to retrofit their airplanes with the redesigned fan cowl structure.

In the final report issued on 13 December 2019 it stated that Boeing was working on a design enhancement 'that would fully address the safety recommendation from the NTSB. Once approved by the FAA, that design change will be implemented in the existing NG fleet.'

C-40A

US Navy, Fleet logistics support aircraft. Certified to operate in an all-passenger configuration for one hundred and twenty-one passengers, an all-cargo variant or a 'combi' configuration which can accommodate up to three cargo pallets and seventy passengers on the main deck. This was the only C-40 version without winglets.

There was a modified C-40A configured to include a distinguished visitor compartment for combatant commanders and communications system operator workstation. The C-40B was designed to be an 'office in the sky' for senior military and government leaders. Communications were paramount aboard the C-40B which provided broadband data/video transmit and receive capability as well as clear and secure voice and data communication. It gave combatant commanders the ability to conduct business anywhere around the world using on-board internet and local area network connections, improved telephones, satellites, television monitors, and facsimile and copy machines. The C-40B also has a computer-based passenger data system.

A further sub-variant of the modified C-40A was to include a convertible cargo area. The aircraft could be converted for medevac, passenger transport or distinguished visitors such as members of the Cabinet and Congress. The C-40C was not equipped with the advanced communications capability of the C-40B. Unique to the C-40C was the capability to change its configuration to accommodate from forty-two to one hundred and eleven passengers.

C-40B
US Air Force, high-priority personnel transport and communications aircraft.

C-40C
Air National Guard and Air Force Reserve Command high-priority personnel transport aircraft.

E-737
Airborne Early Warning and Control (AEW& C)

The 737 Airborne Early Warning and Control was designed for countries that did not need - or could not afford - the capability of the much bigger 767 or 707 AWACS. The base aircraft was essentially a Boeing Business Jet, which has the 737-700 fuselage with the stronger 737-800 wing to support its extra weight and the BBJ aux fuel tanks. The 737 AEW&C cost between $150 million and $190 million, which compared with about $400 million for the 767 AWACS. It carried a mission crew of between six and ten in the forward cabin.

Its main external feature was the 'Top Hat' antenna. This was a phased-array, Multi-role Electronically Scanned Array (MESA) radar sensor developed by Northrop Grumman and mounted in a rectangular fairing over the rear fuselage. The antenna alone weighed 6500 pounds and was 10.7 metres long. It provided fore and aft coverage with a low drag profile which allowed the system to be installed without a significant impact on aircraft performance. The pylon air intakes were for

The E-7A Wedgetail in flight. The highly modified airliner provides Australia with one of the most advanced air battlespace management capabilities in the world. Based at RAAF Base Williamtown, the six E-7A Wedgetails significantly improve the effectiveness of the Australian Defence Force. They are capable of communicating with other aircraft and providing air control from the sky, and can cover four million square kilometres during a single ten -hour mission. *(RAAF)*

The first Peace Eagle for the Türk Hava Kuvvetleri (Turkish Air Force) wearing the test registration N356BJ. *(Boeing)*

the liquid cooling system.

Inevitably, the Top Hat reduced the airflow over the fin and rudder which necessitated the addition of two ventral fins that increased the directional stability for engine-out flight. Also, the elevator feel pitot probes had had to be moved higher up the fin away from the disturbed air flow.

Other modifications included a new section 41 with a cut-out for an air-to-air refuelling receptacle and nose, wingtip and tail mounted counter measure systems. The aircraft will also have chaff and flare dispensers and approx 60 antenna and sensor apertures. The Integrated Drive Generators are uprated to 180kVA and can be seen bulging from the engine cowls.

Dry Operating Weight was expected to be just over 110,000 pounds. Each AEW&C contained an extra 863 electronic boxes, three hundred kilometres of wiring and four million lines of software code more than a standard 737 NG.

The first green aircraft arrived at Wichita in December 2002 for structural modifications. Flight testing of the airframe ran from May 2004 until July 2005 with the aircraft logging more than 500 flight hours in 245 flights.

According to Boeing 'The plane performed superbly in terms of its avionics, structure, systems, flight handling characteristics and performance'. This was followed by flight testing of the mission system, including the MESA radar. All appeared to be going well for the project until 2006 when the first of the delays was announced because of 'development and integration issues with certain hardware and software components'. Deliveries eventually began to Australia in late 2009 but did not reach full operational capability until 2012.

The aircraft is known as the 'Wedgetail' by the Royal Australian Air Force after the Australian Wedgetail Eagle, which '...has extremely acute vision, ranges widely in search of prey, protects its territory without compromise and stays aloft for long periods of time.' The Turkish Air Force call theirs the 'Peace Eagle', presumably for similar reasons.

Chapter Six

The Rudder Mysteries

The safety record of the 737 was good, but in the early 1990s a series of mysterious accidents brought the design into sharp focus, resulting in a vast redesign and retrofit programme that lasted well into the new millennium that was little known about by the general public.

On 3 March 1991, United Airlines flight UA585, a 737-200Adv, crashed on approach to Colorado Springs. The aircraft departed from controlled flight approximately one thousand feet above the ground and struck an open field. After a twenty-one month investigation, the Board issued a report on the crash in December 1992. In that report, the NTSB said it, '...could not identify conclusive evidence to explain the loss of the aircraft', but indicated that the two most likely explanations were either a malfunction of the airliner's directional control system or an encounter with an unusually severe atmospheric disturbance.

Then, on 11 April 1994, at 37,000ft over the Gulf of Honduras, Continental Airlines pilot Ray Miller reported the crew heard a muffled thump and his aircraft rolled violently to the right and continued to pull to the right for another eighteen minutes; the Boeing 737-300 landed safely.

Unlike other twin-engine large transport aircraft, the Boeing 737 was designed with a single rudder panel and single rudder actuator. The single rudder panel was

Dan-Air was another UK charter and scheduled service airline that operated a mix of Boeing products. Here 737-200 'Charlie Victor' is seen during turnaround at Rhodes Airport. *(author)*

controlled by a single hydraulic Power Control Unit (PCU). Inside the PCU was a dual servo valve that, based on input from the pilot's rudder pedals or the aircraft's yaw damper system, directs the flow of hydraulic fluid to move the rudder. The PCU for affected Boeing 737 aircraft was designed by Boeing and manufactured by Parker Hannifin.

Before proceeding further it is worth explaining a few items of technical aspects. In simple terms, at the rear of the fuselage of most aircraft one finds a vertical stabiliser and a rudder. The stabiliser is a fixed-wing section whose job is to provide stability for the aircraft, to keep it flying straight. The vertical stabiliser prevents side-to-side, or yawing, motion of the aircraft nose. The rudder is the small moving section at the rear of the stabiliser that is attached to the fixed sections by hinges. Because the rudder moves, it varies the amount of force generated by the tail surface and is used to generate and control the yawing motion of the aircraft.

The rudder is used to control the position of the nose of the aircraft. In spite of popular belief, it is not used to turn the aircraft in flight. Aircraft turns are caused by banking the aircraft to one side using either ailerons or spoilers. The banking creates an unbalanced side force component of the large wing lift force which causes the aircraft's flight path to curve. The rudder input ensures that the aircraft is properly aligned to the curved flight path during the manoeuvre. Otherwise, the aircraft would encounter additional drag or even a possible adverse yaw condition in which, due to increased drag from the control surfaces, the nose would move farther off the flight path.

The rudder works by changing the effective shape of the airfoil of the vertical stabiliser, changing the angle of deflection

A publicity shot of Piedmont Airlines' N734N, a 737-201. The airliner was delivered on 30 May 1968, and after passing trhrough a number of owners was scrapped in September 1994. Note the 'short' engine nacelles in this picture! *(author's collection)*

G-BGDS B737-236 Adv *River Severn* of British (Airways) as seen at London Heathrow in an early colour scheme. *(author's collection)*

at the rear of an airfoil will change the amount of lift generated by the foil. With increased deflection, the lift will increase in the opposite direction. The rudder and vertical stabiliser are mounted so that they will produce forces from side to side, not up and down.

On the 737, yaw control was achieved by a single graphite-composite rudder panel. A single rudder Power Control Unit (PCU) controls rudder panel deflection. A standby rudder PCU provides back up in the event of malfunction of the main rudder PCU. There is no manual reversion for yaw control. The only internal indication of rudder panel defection was the pedal position, which always accurately reflects control surface deflection. The total authority of the control surface was modulated in relation to aircraft's indicated airspeed using 'blowdown', - that being a constant pressure applied to the surface by the actuator, and the movement of the panel reduces accordingly as the dynamic pressure on it increases. For this reason,

A full power during take-off the P&W JT-8Ds could be 'smokey Joes'. Here an Aer Lingus 737-200, almost certainly EI-ASH, gets airborne. *(Dee Diddley collection)*

Frontier Airlines was a American airline formed in 1950 by a merger of Arizona Airways, Challenger Airlines, and Monarch Airlines. Headquartered at Stapleton Airport in Denver, Colorado, the airline ceased operations in 1986. Here their 737-291 N7348F is seen taxiing out for another service from Columbus, Ohio in 1980. *(author)*

the maximum rudder pedal movement was reduced with increasing airspeed. Maximum rudder panel deflection being approximately +/-15 degrees on the ground, reducing to around +/-8 degrees at a typical cruise altitude.

The rudder PCU consisted of an input shaft/crank mechanism, a dual concentric servo valve to control porting the fluid to the rudder actuator, and a yaw damper actuator. The rudder actuator is a tandem actuator, having two internal piston areas for each hydraulic source (A & B). The actuator was capable of positioning the rudder panel with either one or both main hydraulic sources available, though with one source inoperative a reduced rudder panel deflection would result due to blowdown at higher airspeeds.

Flow of hydraulic fluid to the rudder actuator is controlled by the dual servo-valve. This is a complex dual concentric cylinder with an outer and inner slide. During normal rudder pedal inputs, sufficient rudder panel deflection was catered for by the primary, or inner valve alone. However, should a larger panel deflection be required or a higher rate rudder input be commanded, the secondary, or outer sleeve moves in addition to allow extra fluid to the actuator. Movement of the outer sleeve was typically no more than one millimetre. Position of both sleeves of the servo valve was controlled by a complex mechanism of bell cranks, input rod and summing lever, the geometry of which was such as to provide movement of the sleeves in relation to the body of the valve.

The yaw damper, incorporated to prevent Dutch roll - a type of aircraft motion consisting of an out-of-phase combination of 'tail-wagging' - termed 'yaw' - and rocking from side to side, termed 'roll'. Dutch roll stability could be artificially increased by the installation of a yaw damper. It is connected in parallel with the main servo valve and includes its own actuator, powered by hydraulic system B. This actuator applies its own input to the input shaft/crank mechanism to bring about a movement of the servo-valve and hence a rudder panel deflection. No pedal movement results from yaw damper operation. The total authority of the yaw damper is approximately +/-2.5 degrees.

After the 11 April 1994 incident, Continental removed the flight data recorder and rudder Power Control Unit from the incident aircraft and provided them to Boeing for investigation.

Boeing concluded that hydraulic fluid had leaked from the PCU onto the yaw-damper signalling component, creating an open electrical circuit that inadvertently moved the rudder two and a half degrees to the left. Such a deflection should have caused the jet to veer only slightly off course, something easily controllable by the pilots. Moreover, Boeing said the problem could not have lasted more than 110 seconds, a finding that was disputed by Captain Miller.

Then on 8 September 1994, N513AU, a 737-300, was operating as US Air flight US427, a scheduled flight from Chicago O'Hare International Airport to Pittsburgh International Airport, with a final destination of West Palm Beach, Florida.

The flight crew were Captain Peter Germano, 45 and First Officer Charles B 'Chuck' Emmett III, 38. The Flight attendants were Stanley Canty and Sarah Slocum-Hamley.

When approaching Pittsburgh runway 28R Air Traffic Control reported traffic in the area, which was confirmed in sight by the First Officer. At that moment, the aircraft was levelling off at 6000 feet while flying at a speed of 190 knots and rolling out of a fifteen degree left turn with a roll rate of two degrees per second. The flaps were at 1, the undercarriage still retracted and autopilot and auto-throttle systems engaged.

Post-crash analysis of the Flight Data Recorder (FDR) revealed that the aircraft suddenly entered the wake vortex of a Delta Airlines Boeing 727 that was seventy seconds, or 4.2 nautical miles in front of it. Over the next three seconds, the aircraft rolled left to eighteen degrees of bank. The autopilot attempted to initiate a roll back to the right as the aircraft went in and out of a wake vortex core, resulting in two loud 'thumps'. The First Officer then manually overrode the autopilot, without disengaging it, by putting in a large right-wheel command at a rate of one hundred and fifty degrees per second. The airliner started rolling back to the right at an

Societé Anonyme Belge d'Exploitation de la Navigation Aérienne, meaning Belgian Corporation for Air Navigation Services was better known internationally by the acronym SABENA. Here OO-SDC is seen with the camshell thrust reverser buckets deployed at London Heathrow. *(author)*

Maersk Air A/S was a Danish airline which operated between 1969 and 2005. Owned by the A. P. Møller–Mærsk Group, it operated a mix of scheduled and chartered passenger and cargo services. Headquartered at Dragør, its main operating bases were Copenhagen Airport, Billund Airport and Esbjerg Airport. Boeing 737-2L9 was obtained new in 1976 *(Maersk Air A/S)*.

acceleration that peaked at thirty-six degrees per second, but the aircraft never reached a wings-level attitude.

At 19.03:01 the aircraft's heading slewed suddenly and dramatically to the left with full left rudder deflection. Within a second of the yaw onset, the roll attitude suddenly began to increase to the left, reaching thirty degrees. The aircraft pitched down, continuing to roll through fifty-five degrees of left bank. At 19.03:07, the pitch attitude approached minus twenty degrees; the left bank increased to seventy degrees, and the descent rate reached 3600 feet per minute. At this point, the aircraft stalled. Left roll and yaw continued, and the aircraft rolled through inverted flight as the nose reached ninety degrees down, approximately 3600 feet above the ground. The 737 continued to roll, but the nose began to rise. At 2000 feet above the ground, the aircraft's attitude passed forty degrees nose low, and fifteen degrees left bank. The left roll hesitated briefly but continued, and the nose again dropped. The airliner descended fast and the roller-coaster ride ended when the aircraft impacted the ground nose first at 261 knots in an eighty-degree nose down, sixty degrees left bank attitude and with significant sideslip. All 132 on board were killed.

The NTSB realised the crash of Flight 427 might have been caused by an unintended or uncommanded rudder movement, similar to the suspected - but not yet established - cause of the Flight 585 crash. As a result, the NTSB conducted additional testing on United Flight 585's PCU servo during its Flight 427 investigation.

One of the problems facing

ZK-NAR was a 737-219. *(Air New Zealand).*

investigators was the relative lack of precision in the data produced by the FDR, which only recorded control inputs at periodic intervals with significant time gaps between samples, gaps during which no data was recorded no matter what the pilot did with the controls. This lack of precision led to it being possible for Boeing to interpret the data differently from the way the NTSB did, leading the manufacturer to suspect and insist that the pilot had responded incorrectly to a wake turbulence incident.

At the request of the NTSB, data from the Penny & Giles quick access recorder - 'QAR' - of a British Airways (BA) Boeing 747-400 London to Bangkok flight in which the aircraft had suffered an uncommanded elevator movement and momentary elevator reversal on take-off, the aircraft then continuing its flight and landing safely, was supplied to the NTSB by British Airways. Unlike a standard FDR, the QAR sampled control input data at much shorter time intervals, as well as sampling and recording many more other

Chrono Aviation was a Canadian charter airline based in Montreal, Quebec City and Rimouski They offered the former New Zealand 737-219C as C-GTVO in either executive, cargo or combi configuration that was equipped for full unpaved strip operation. *(Chrono Aviation)*

aircraft parameters. This British Airways data led to renewed suspicion of the similar valve design used on the 737 rudder. As a result of this British Airways incident, Boeing had, in fact, modified the design of the 747 elevator servo system, and the modified system had been retroactively fitted to all 747-400s in service.

At the time a number of conflicting views were held into the incidents involving the two early accidents

The US Air View: *'The probable cause of this accident was an uncommanded, full rudder deflection or rudder reversal that placed the aircraft in a flight regime from which recovery was not possible using known recovery procedures. A contributing cause of this accident was the manufacturer's failure to advise operators that there was a speed below which the aircraft's lateral control authority was insufficient to counteract a full rudder deflection.'*

The ALPA View: *'Based on the evidence developed during the course of this accident investigation, ALPA believes that the airplane experienced an uncommanded full rudder deflection. This deflection was a result of a main rudder power control unit (PCU) secondary valve jam which resulted in a primary valve overstroke. This secondary valve jam and primary valve overstroke caused USAir 427 to roll uncontrollably and dive into the ground. Once the full rudder hardover occurred, the flight crew was unable to counter the resulting roll with aileron because the B737 does not have sufficient lateral control authority to balance a full rudder input in certain areas of the flight envelope.'*

The Boeing View: Charlie Higgins, Boeing vice president, Airplane Safety & Performance, said that, *'the rudder PCUs from the 737s that crashed near Colorado Springs in 1991 and Pittsburgh in 1994 were both thoroughly examined as a part of the NTSB's accident investigations. No such jam was detected in either unit.'*

Mike Denton, former chief engineer for the 737 said, *'We don't know if there was an airplane system failure; we don't know if the flight crews had their foot on the pedal the full time.'*

Conclusions from Boeing's report to the NTSB, 30 Sep 1997: Several elements leading to this accident are clear:

1. The crew was startled by the severity of an unexpected wake vortex encounter.
2. A full rudder deflection occurred.

A colour scheme that always looked good on the 737 was Delta Airlines, which operated a large number of various sub-types. Here N375DL is seen on approach to Cincinnati/Northern Kentucky International Airport in 1996. *(author)*

Olympic Airways operated a mixed fleet of 737s - this publicity picture was taken just after the opening of the new Athens Eleftherios Venizelos International Airport. *(Olympic Airways via author)*

However, the events that led to the full rudder deflection are not so clear:

- There is no certain proof of airplane-caused full rudder deflection during the accident sequence. The previously unknown failure conditions that have been discovered in the 737 rudder PCU have been shown to not be applicable to Flight 427 or any other conditions experienced in commercial service.
- There is no certain proof that the flight crew was responsible for the sustained full left rudder deflection. However, a plausible explanation for a crew-generated left rudder input must be considered, especially given the lack of evidence for an airplane-induced rudder deflection.
- In Boeing's view, under the standards developed by the NTSB, there is insufficient evidence to reach a conclusion as to the probable cause of the rudder deflection.

3. The airplane entered a stall and remained stalled for approximately 14 seconds and 4,300 feet of altitude loss.

The NTSB and FAA View

NTSB investigator Malcolm Brenner said in explaining the crash of Flight 427: '*The pilots were trying to deal with emergencies with reasonable actions but could not understand what was happening in the time available*'

Dennis Crider, chairman of the NTSB's Aircraft Performance Group, told the board members: 'A rudder reversal scenario will match all three events'.

The FAA View: The FAA argued that no one will ever know the cause with any degree of certainty, so it has focused on making the aircraft safer. Over the next four years nine specific Airworthiness Directives were issued to the operators.

3 March 1994 - AD 94-01-07: '*Within 750 flight hours Perform a test of the main rudder PCU to detect internal leakage of hydraulic fluid. Repeat at 750hr intervals unless replaced with new main rudder PCU.*'

27 November 1996 - AD 96-23-51: *'Within ten days perform a test to verify proper operation of the rudder PCU.'*

17 January 1997 - AD 96-26-07: *This introduced, within 30 days, a new QRH recall procedure called 'UNCOMMANDED YAW OR ROLL' and changed the 'JAMMED FLIGHT CONTROLS' procedure to include a section entitled 'JAMMED OR RESTRICTED RUDDER'.*

Note: Following the NTSB report of 16 April 1999, these procedures were replaced by *'UNCOMMANDED RUDDER'* and *'UNCOMMANDED YAW OR ROLL' in AD 2000-22-02.*

19 March 1997 - AD 97-05-10: *'Within ninety days inspect the internal summing lever assembly of the main rudder PCU.'*

9 June 1997 - AD 97-09-15: *'Within five years or 15,000 flight hours Requires a one-time inspection of the engage solenoid valve of the yaw damper to determine the part number of the valve, and replace if necessary.'*

1 August 1997 - AD 97-14-03: *'Within three years install a newly designed rudder-limiting device that reduces the rudder authority at flight conditions where full rudder authority is not required (a rudder pressure reducer (RPR)). . . .Install a newly designed yaw damper system that improves the reliability and fault monitoring capability.'*

4 August 1997 - AD 97-14-04: *'Within two years Requires replacement of all rudder PCUs with a newly designed main rudder PCU.' And 'Replace the vernier control rod bolts with newly designed vernier control rod bolts.'*

20 January 1998 - AD 97-26-01: *'Within eighteen months or 4,500 hours Perform an inspection to detect galling on the input shaft and bearing of the standby rudder PCU. .. and replace the input bearing of the standby rudder PCU with an improved bearing.'*

17 February 1998 - AD 98-02-01: *'Within 3,000 hours Remove the yaw*

Going about Combi business! Royal Brunei Airlines 737-2M6 seen during turnaround, with cargo going in the big door up front and passengers about to board the small door at the back. *(author)*

Frontier Airlines' 737-238 N234TR on finals into Portland International Airport. This iteration - with no similarity to the earlier version of Frontier Airlines apart from the name and the fact that they operated 737-200s began operations in 1994. Based in Denver, they were an early user of artwork on the vertical, as seen here. *(author)*

damper coupler, replace the internal rate gyroscope with a new or overhauled unit, and perform a test to verify the integrity of the yaw damper coupler.'

Not quite two years later, on 9 June 1996, while operating a passenger flight the crew of N221US, an Eastwind Airlines Flight 517 temporarily lost control of their Boeing 737-200 due to a rudder malfunction.

Flight 517 was a scheduled Eastwind Airlines passenger flight from Trenton-Mercer Airport in Trenton, New Jersey, to Richmond International Airport in Richmond, Virginia. On the flightdeck was Captain Brian Bishop and First Officer Spencer Griffin; there was a total of 53 people were on board.

Flight 517 departed Trenton without incident and encountered no turbulence or unusual weather en route to Richmond. While on approach to Richmond International Airport, at an altitude of about 5,000 feet the Captain felt a brief 'kick' on the right rudder pedal. Around the same time, a flight attendant at the rear

'Super 737' G-BKYA. This aircraft was the only one to wear these experimental metal roof colours and was delivered from Boeing like this but the colour was not adopted by BA which choose 'Landor' colours instead. *(author's collection)*

of the plane heard a thumping noise underneath her. As the airliner continued to descend through 4,000 feet, the Captain suddenly experienced a loss of rudder control and the plane rolled sharply to the right.

Attempting to regain control, the Captain tried to apply full left rudder, but the rudder controls were stiff and did not respond to his commands. The Captain applied left aileron and increased power to the right engine to try to stop the roll. The 737 temporarily stabilised, and then rolled to the right again. The crew performed their emergency checklist and attempted to regain control of the aircraft, and after several seconds they abruptly regained control. The airliner operated normally for the remaining duration of the flight. No damage occurred to the aircraft as a result of the incident. One flight attendant suffered minor injuries. No other passengers or crew aboard Flight 517 were injured.

The NTSB investigated the incident, with a particular focus on determining whether the events of Flight 517 were related to previous Boeing 737 crashes. During the investigation, the NTSB found that prior to the 9 June incident, flight crews had reported a series of rudder-related events on the incident aircraft, including abnormal 'bumps' on the rudder pedals and uncommanded movement of the rudder.

Investigators interviewed the pilots of Flight 517, and removed rudder components from the aircraft for examination, which helped to establish the cause of the previous crashes of United Flight 585 and USAir Flight 427. The NTSB determined that all three incidents could only be explained by pilot error or a malfunction of the rudder system, and based partly on post-accident interviews with the Flight 517 pilots, concluded that rudder malfunctions were likely to have caused all three incidents.

The NTSB also determined that, unlike the United or USAir accidents, the rudder problem on Flight 517 occurred earlier in the landing process and at a higher speed, which increased airflow over the other control surfaces of the aircraft, allowing the pilots to overcome the rudder-induced roll.

Because Eastwind Flight 517 had landed safely, the NTSB was also able to

N284AU of US Air - a 737-2B4 that was built in 1984, and scrapped in 2013 after passing through five operators. *(Hugh Jampton Collection)*

A pocket-rocket gets airborne! Malev's 737-2M8 HA-LEB was previously operated by TEA Belgium, but is seen here leased from Guiness Peat Aviation. *(author)*

perform tests on the actual aircraft that had experienced problems. In addition, because the pilots of Flight 517 had survived, the NTSB was able to interview them and gain additional information on their experience. The flight's captain told the NTSB in a post-incident interview that they had not encountered any turbulence during the flight, and that, during their landing descent, he felt the rudder 'kick' or 'bump' even though neither pilot had moved the rudder pedals. When the amended accident report for Flight 585 that found the same probable cause for that accident as well.

Other incidents involving the 737 were also suspected as being caused by rudder PCU issues. On 6 June 1992, 737-204 Adv, HP-1205CMP of COPA Airlines, operating as Flight 201 took off from runway 21L at Tocumen International Airport in Panama City at 8:37 p.m local time.

The flight crew was Captain Rafael Carlos Chial, 53, and First Officer Cesareo Tejada. The Flight Attendants were Iris Karamañites, Flor Díaz, Vanessa Lewis, Xenia Guzmán, and Ramón Bouche. They were on a scheduled passenger flight to Cali, Colombia At 8:47 p.m. about ten minutes after take-off, Captain Chial contacted Panama City Air Traffic Control, requesting weather information. The controller reported that there was an area of very bad weather 30–50 miles from their position.

N463GB, a 737-293 of Braniff, in a basic American Airlines colour scheme. *(Kirk Smeeton Collection)*

excel airways was established in 1994 as Sabre Airways, and started operations on 17 December 1994. The name excel was adopted in November 2000. It ceased operations when it went into administration on 12 September 2008. G-EXAE, a 737-8Q8 is seen here taxiing in at Samos Aristarchos International Airport, Greece. *(author)*

At 8:48 p.m. Captain Chial made another radio contact requesting permission from Panama City ATC to fly a different route due to the severe weather ahead. The new route would take the airliner over Darién Province. Six minutes later, at 8:54 p.m. Panama City Control Center received a third message from Captain Chial, who reported problems with the aircraft and made a request to turn back to Tocumen, which was granted.

However, at 8:56 p.m while flying at an altitude of 25,000 feet, Flight 201 entered a steep dive at an angle of eighty degrees to the right and began to roll uncontrollably while accelerating towards the ground. Despite the attempts by Captain Chial and First Officer Tejada to level off, the airliner continued its steep dive, until it exceeded the speed of sound and started to break apart at 10,000 feet. Flight 201 crashed into a jungle area within the Darien Gap at 486 knots, instantly killing everyone on board.

Finally, there was SilkAir Flight 185 flown by 9V-TRF, a 737-300 a scheduled SilkAir passenger flight from Jakarta, Indonesia, to Singapore, that crashed into the Musi River near Palembang in southern Sumatra, on 19 December 1997.

Carrying 97 passengers and a crew of seven, the Boeing 737 departed Jakarta's Soekarno-Hatta International Airport's runway 25R at 15:37 local time for a planned eighty-minute flight to Singapore Changi Airport, with Captain Tsu Way Ming 41, of Singapore, a former A-4 Skyhawk pilot, at the controls along with First Officer Duncan Ward, 23, of New Zealand. Generally, fair weather was expected for the route, except for some thunderstorms near Singkep Island, seventy-five miles south of Singapore.

The aircraft was cleared to climb to flight level 350 and to head directly to Palembang. At 15:47:06, while climbing through 24,500 feet, the crew requested clearance to proceed directly to waypoint PARDI (0°34′S 104°13′E). At 15:53, the crew reported reaching the cruise altitude of FL350 and was cleared to proceed directly to PARDI, and to report abeam Palembang. The cockpit voice recorder (CVR) ceased recording at 16:05.

Flight 185 remained level at FL350 until it started a rapid and nearly vertical dive around 16:12. While descending through 12,000 feet, parts of the aircraft, including a great extent of the tail section, started to separate from the aircraft's fuselage due to high forces arising from the nearly supersonic dive. Seconds later, the aircraft impacted the Musi River, near Palembang, Sumatra, killing all 104 people on board. The time it took the aircraft to dive from cruise altitude to the river was less than one minute. The plane was travelling faster than the speed of sound for a few seconds before impact. Parts of the wreckage were embedded fifteen feet into the riverbed.

The aircraft broke into pieces before impact, with the debris spread over several kilometres, though most of the wreckage was concentrated in a single 200 feet by 260 feet area at the river bottom. No complete body, body part or limb was found, as the entire aircraft and passengers disintegrated upon impact.

The cause of the crash was independently investigated by two agencies in two countries: the United States National Transportation Safety Board (NTSB) and the Indonesian NTSC.. The NTSB, which had jurisdiction based on Boeing's manufacture of the aircraft in the USA, investigated the crash under lead investigator Greg Feith. Its investigation concluded that the crash was the result of deliberate flight control inputs, most likely by the captain. The Indonesian NTSC, led by Engineering Professor Oetarjo Diran, was unable to determine a cause of the crash.

Another potential factor that led to the crash of the 737 aircraft was the power control unit (PCU) that controlled the aircraft's rudder. Although the NTSB and PCU manufacturer Parker Hannifin had already determined that the PCU was

C-GCPS was a 737-217 and is seen here in a hybrid CP-Air and Canadian Airlines colour scheme, photographed around the time the two airlines merged in 1987. *(author's collection)*

With the fleet number '750' on the nosewheel door and tip of the fin, C-GMPW, formerly of Pacific Western comes in to land. One of a fleet of around fifty series 200s operated by Canadian Airlines International this is one of two that carried 'Canadian Holidays' on one side and 'Vacances Canadien' on the other. *(author)*

properly working, and thus not the cause of the crash, a private investigation into the crash for a civil lawsuit tried by jury in a state court in Los Angeles, which was not allowed to hear or consider the NTSB's and Parker Hannifin's conclusions, decided that the crash was caused by a defective servo valve inside the PCU, based on forensic findings from an electron microscope which determined that minute defects within the PCU had caused the rudder hard-over and a subsequent uncontrollable flight and crash. The manufacturer of the aircraft's rudder controls and the families later reached an out-of-court settlement.

Eventually, as a result of the NTSB's findings into all the incidents, the Federal Aviation Administration ordered that the servo valves be replaced on all 737s. The FAA also ordered new training protocols for pilots to handle unexpected movement of flight controls.

So began a discreet and worldwide process that went on for a number of years.

Everyone thought that the troubles were over - then two further rudder incidents in 1999 caused concern because they

Wien Air Alaska's 737-210C N493WC fitted with a partial gravel runway kit; the engine vortex dissapator is visible, but the nose-ski appears to have been removed. *(Liam Cannuck collection)*

The distinctive later scheme worn on Aer Lingus aircraft. EI-ASA was a 737-248 and was regularly seen on the Dublin - Heathrow - Dublin route. *(author's collection)*

involved aircraft retrofitted with a PCU redesigned to make a valve jam impossible. After the incidents, the dual-servo valves on both airliners were inspected for cracks but none were found.

On 19 February 1999 the pilots of the United Airlines Boeing 737-300 reported sluggish rudder control during a ground check while taxiing at Seattle-Tacoma International Airport. The NTSB said the apparent problem was a mispositioned valve-spring guide in the rudder's power-control unit.

Then, a few days later on 23 February 1999, a Metrojet 737-200 made a precautionary landing in Baltimore after the aircraft rolled slightly and changed direction during cruise flight. The aircraft's rudder moved involuntarily at two rates, first slowly and then more rapidly over to the point where the rudder deflects to its maximum extent, known as a 'hardover.' The pilots could not make the rudder move, including using the emergency procedures of turning off the autopilot and the yaw damper. The pilots reported that when they turned off the aircraft hydraulic pressure, the rudder 'snapped back' into position but continued to 'chatter' and vibrate through the emergency landing. No known scenario could cause such an event. The NTSB examined the PCU but did not find evidence of a cause. All that investigators were able to conclude is that a rudder deflection did occur, according to information from the airliner's flight-data recorder.

During March 1999 the NTSB released a report stating that 'although there was no hard physical evidence, both the Colorado and Pittsburgh crashes were probably caused by an abrupt rudder movement that surprised the crew and sent the planes spiralling into an uncontrollable dive.'

Then, on 16 April, the NTSB made the following recommendations to the FAA:

• All existing and future Boeing 737s and future transport category aircraft must have a reliably redundant rudder actuation system.

• Convene an engineering test and evaluation board (ETEB) to conduct a failure analysis to identify potential failure modes. The board's work should be completed by March 31, 2000, and published by the FAA.

• Amend the FARs to require that transport-category airplanes be shown to be capable of continued safe flight and landing after jamming of a flight control at any deflection possible, up to and including its full deflection, unless such a jam is shown to be extremely improbable.

• Revise AD 96-26-07 so that procedures for addressing a jammed or restricted rudder do not rely on the pilots' ability to center the rudder pedals as an indication that the rudder malfunction has been successfully resolved.

• Require all operators of the Boeing 737 to provide their flight crews with initial and recurrent flight simulator training in the 'Uncommanded Yaw or Roll' and 'Jammed or Restricted Rudder' procedures. The training should demonstrate the inability to control the airplane at some speeds and configurations by using the roll controls (the crossover airspeed phenomenon) and include performance of both procedures in their entirety.

• Require Boeing to update its Boeing 737 simulator package to reflect flight test data on crossover airspeed and then require all operators of the Boeing 737 to incorporate these changes in their simulators.

• Evaluate the Boeing 737's block manoeuvring speed schedule to ensure the adequacy of airspeed margins above crossover airspeed for each flap configuration, and revise block manoeuvring speeds accordingly.

• Require that all Boeing 737s be equipped, by July 31, 2000 with an FDR which records the minimum

Canadian North Inc. is a wholly Inuit-owned airline headquartered in Kanata, Ontario. It operates scheduled passenger services to communities in the Northwest Territories, Nunavik and Nunavut. Southern gateways include Edmonton, Montreal and Ottawa. Here C-GOPW, one of their 737-275 Combis loads passengers through the rear door at Ottawa International Airport. *(Liam Cannuck collection)*

A pair of London Stansted residents from the 1990s. Left is G-IGOI, a 737-33A of low-cost carrier go, and below is G-UKLA of Air UK Leisure, a 737-4Y0. *(author's collection)*

parameters applicable to that airplane, plus the following parameters: pitch trim; trailing edge and leading edge flaps; thrust reverser position (each engine); yaw damper command; yaw damper on/off discrete; standby rudder on/off discrete; and control wheel, control column, and rudder pedal forces (with yaw damper command; yaw damper on/off discrete; and control wheel, control column, and rudder pedal forces sampled at a minimum rate of twice per second).

The Flight Control Engineering and Test Evaluation Board (ETEB) was formed by the FAA in May 1999 to take a fresh look at the 737 rudder in the most in-depth scientific study ever of any commercial aircraft system. FAA officials stressed that the board was formed only to determine what could happen, not to evaluate the probability that it would happen. And they cautioned that they had not found anything that they believed would be sufficiently probable to warrant grounding the aircraft or even ordering immediate design changes. Any eventual design change would be required on all existing and newly manufactured 737s.

The FAA used two criteria to assemble the ETEB members, technical expertise and no connection with, or knowledge of, the 737 rudder system. John McGraw, manager of the FAA's Airplane and Flight Crew Interface Branch in Seattle, headed the board, which was composed of scientists from the FAA, NASA, the Defence Department, ALPA, the Air Transport Association, the Russian Air Transport Accident Investigation Commission, Ford Motor Co. and Boeing. However, Boeing personnel came from Boeing Military and Boeing Long Beach, not from Boeing Seattle where the 737 was made because Seattle engineers could be too familiar with the rudder.

Dr Davor Hrovat, an engineer from Ford was included on the team because he had developed a way to use ultrasound to determine movements of internal parts, such as the inner slide of the 737's dual concentric servo valve. A high-level, seven-person 'challenge team' of outside experts was also formed to review every step of the board's work. It included Col. Charles Bergman, the Air Force deputy chief of safety; Vladimir Kofman, chairman of the Russian accident investigation commission; and Tom Haueter, chief of the major investigations division of the NTSB.

The ETEB had full access to everything it needed, including a special test aircraft owned by Purdue University and laboratory space at Boeing. They even constructed a first-of-a-kind test device called a 'fin rig', which was a full 737 vertical tail fin and rudder connected to an aircraft engineering simulator. Any rudder control commands by a pilot were mimicked in real-time on the rudder, which was placed where the pilot could easily see its movements.

Engineers used a vibration table to give a good shaking to rudder components, a 'cold box' that could produce realistic flight temperatures, and a device that could produce sudden spikes in hydraulic fluid temperatures from 65 degrees below zero to 210 degrees above zero. They could also spray water into the rudder mechanism, producing ice. It was during these ice tests that the engineers found a new and unsuspected failure scenario.

During the study, the ETEB brought in ten flight crews from four airlines on 737s to fly the fin rig simulator. They used the existing recovery procedures to deal with

About to do a 180 on the frying pan at the 35 end of the runway of Corfu International Airport 'Ioannis Kapodistrias' is Smartwings 737-82R(W) OK-TSR. This airport, on one of the most popular of the Greek islands, can almost be called a mecca for independant airlines in the summer months, bringing 737s from all over Europe, and often further afield. It also has a causeway running close to one end of the runway, and a number of hotels - including the Royal Boutique, room 420 - that has stunning views of the runway from their balconies! *(author)*

A typical day at London Heathrow in 1979. 747s, TriStars and Tridents. Closest to the camera is PH-TVH, a 737-222 on lease from the Dutch airline Transavia. *(Kirk Smeaton Collection)*

about forty different rudder failure modes. They found, as expected, that any rudder hardover, while taking off or landing, moving slowly and at low altitude, would be catastrophic. And they found that these pilots, who fly the 737s routinely for airlines and had normal training, performed poorly in trouble-shooting rudder problems.

While these investigations were going on a further Airworthiness Direction was issued: 28 Jun 1999 (Revised 7 Dec 1999) - AD 99-11-05: 'Within 16 months perform repetitive displacement tests of the secondary slide in the dual concentric servo valve of the PCU to detect cracks in a joint in the servo valve that regulates the intake of hydraulic fluid to the PCU.'

In October 1999 the NTSB adopted a revised final report on the UA585 and US427 crashes. The Board said that the most likely cause of the accident was the movement of the rudder to its limit in the direction opposite that commanded by the flight crew, 'most likely' because of a jam in the device that moves the rudder. The decision tracks information learned from the investigation of UA5 85, US427 and the Eastwind incident.

A preliminary draft report was issued by the ETEB on 12 April 2000 in which they found that the 'JAMMED OR RESTRICTED RUDDER' procedures formulated by Boeing and often modified by airlines were 'confusing and time-consuming'. They said the pilots showed a lack of training and situational awareness in controlling malfunctions, and as they prepared to land, they never checked to be sure the rudder was operating properly.

The evaluation board detected thirty failures and jams that could be catastrophic on take-off and landing. However, because aircraft travel faster at higher altitudes and other control systems could overcome the force of a rudder, it considered no failures at cruising altitude to be catastrophic. Nonetheless, sixteen of those failures and jams at higher altitudes would be 'hazardous', meaning they would require

OO-RVM, a 737-2Q8 of Air Belgium. *(Hugh Jampton collection)*

prompt pilot action to prevent a crash.

The report said that another twenty-two 'latent failures,' such as a cracked part, combined with single failures and jams, could cause catastrophic or hazardous failures. All but three of the failures found by the panel were also present in the new generation 737-6/7/800. 'The large numbers of single failures and jams and latent failure combinations are of concern.' They also found that current maintenance procedures were insufficient to find hidden problems in the rudder system.

The evaluation board's most unexpected finding was that an ice buildup could cause a 737 rudder to malfunction. The pilot's rudder pedals were connected by cables to a linkage in the tail section. A hydraulic servo valve in the linkage powered an actuator that moved the rudder. The linkage included a summing lever that stopped the rudder at the position specified by the pilot. Mechanical stops prevented the summing lever from moving too far, which prevented proper operation of other levers that shut off hydraulic fluid flowing through the dual control valve, allowing fluid to keep pumping until the rudder went to its maximum deflection.

It is here that the term 'summing lever' needs to be explained. It is also known as a 'free' lever, and any equations are

Former Lufthansa 737-130, N417PE is seen in basic People Express scheme, but with Continental markings, following the merger. *(Matt Black Collection)*

Hooters Air was established in 2003. It was founded and owned by Hooters of America restaurant owner Robert Brooks, who acquired Pace Airlines in December 2002. All flights were operated by Pace Airlines. Aside from its unorthodox neighborhood chain-restaurant tie-in, Hooters Air sought to differentiate itself from other carriers with a distinctive style of in-flight service. The carrier was marketed towards golfers in an effort to bring casual and tournament players to Myrtle Beach's 100+ championship golf courses. Two 'Hooters Girls', dressed in their skimpy restaurant uniforms, were on each flight assisting the traditionally attired in-flight crews with hospitality duties. The company advertised nonstop flights for most routes, using humorous slogans like 'Fly a mile high with us.' All aircraft were painted in Hooters' orange and white company colours featuring the company logo, and mascot 'Hootie the Owl', on the vertical stabiliser. Here 737-2K5 N250TR is seen at Orlando, Florida. *(Hugh Jampton collection)*

derived with the assumption of small angle deviation from initial position. The term comes from the relationship between forces and displacements by summing (or adding) the torques around the fulcrum

The linkage was not pressurised or heated and operated in temperatures as low as -60C. The board found that ice could form in the linkage, jamming the summing lever. Without the equalising force of the lever, the servo valve could continue providing hydraulic pressure; the rudder then would keep moving as far as it could go in the requested direction, a condition known as a hardover. There was no proof that this malfunction had ever occurred in flight because ice would melt afterwards, leaving no marks. FAA officials stressed

Midway was founded on 6 August 1976 and was intended to breathe new life into Midway International Airport, then called Chicago Midway Airport, which had lost most of its scheduled flights to O'Hare International Airport. *(Matt Black Collection)*

A relatively quiet time at London Gatwick, as Norwegian Air Shuttle's 737-86J(W) LN-NIB is pushed back for the start of another service. Norwegian went through a phase of depicting famous Norwegians, such as Helmer Hanseen, the polar explorer, as seen on India Bravo here *(author)*

that this phenomenon had not at the time been tested in flight, but they were nonetheless working on a fix to make certain ice did not form or was cleared away naturally by the movement of the mechanisms.

The evaluation board recommended in the draft that Boeing modify the 737 rudder control system so that 'no single failure, single jam, or any latent failure in combination with any single jam or failure will cause Class I - that is catastrophic - effects'.

In the meantime, the draft report recommended alerting flight crews about early signs of rudder malfunctions, most often rudder 'kicks' that pilots might attribute to yaw damper problems. It also recommended new maintenance inspection procedures, a new cockpit instrument to tell the pilots exactly how the rudder was moving and a new hydraulic system design to allow hydraulic pressure to be cut off to the rudder without affecting other aircraft systems.

On 13 September 2000 the FAA reached an agreement with Boeing to redesign the rudder. Once the directive was issued, the company would have about five years to make the changes in aircraft then flying and be incorporated in all new 737s. Because the redesign could take years to implement, the FAA implemented new training procedures for pilots to use in the event of rudder problems.

Earlier changes in the design had fixed problems with some control mechanisms and an earlier set of emergency rudder control problem training procedures for pilots was put in place, but these were found by the ETEB to be too complex, and pilots had not received enough training to handle them effectively.

The NTSB's chairman, Jim Hall, went on the record to say that he was pleased Boeing and the FAA had finally agreed that there was a need to redesign the 737 rudder control system. The then-current design, Hall said, 'represents an unacceptable risk to the travelling public'. He went on, 'I hope this redesign and retrofit can be accomplished expeditiously so that the major recommendation of our accident report last year will be realised, a reliably redundant rudder system for Boeing 737s'.

The FAA stated that even more had to be done to assure the redundancy of the rudder's safety system and to assure that it was impossible for it to malfunction. Pilots must have more training to handle a rollover caused by a malfunctioning rudder. Boeing said that the new rudder would take three years to develop and be fitted to the first aircraft and the FAA said the last would not be fitted until 2008 because the retrofit work is expected to take as much as two hundred hours per aircraft. Boeing would pay the full cost of the retrofit, estimated at $240 million.

Boeing's Allen Bailey, the engineer in charge of 737 safety certification, said, 'We are not fixing a safety problem with this enhancement we are making'.

The next day NTSB Chairman Jim Hall made a statement on the FAA release of the ETEB rudder study: 'The men and women of the Engineering Test and Evaluation Board can be justifiably proud of the work they have done over the past year and for the final report issued today. The ETEB - made up of representatives from the

It became a trend at the turn of the century for many US passenger terminals to cover their windows with a film that had printed 'dots' in it to reduce the sunlight and heat coming through the glass - something that made it a nightmare to get decent photographs! Seen in May 2018 through one such set of windows in the departure area of Detroit's Metropolitan Wayne County Airport is AeroMexico 737-752W N788XA . *(author)*

The former DeutscheBA 737-36Q D-ADIA, now registered YL-BBJ for airBaltic, seen landing at Paris Charles de Gaulle. *(Robin Banks)*

Federal Aviation Administration and aircraft manufacturers - was the result of a Safety Board recommendation following our investigation of the crash of USAir Flight 427. I think I can speak for my NTSB colleagues by saying that we are gratified that the ETEB essentially confirmed our findings in that accident report. The major finding of both reports is that the Boeing 737 rudder control system has numerous potential failure modes that represent an unacceptable risk to the travelling public. The ETEB found dozens of single failures and jams and latent failures in the 737 rudder system, in addition to the single point of failure we identified in our accident report, that can result in the loss of control of the airplane. Although the failure mechanism that we believed led to the crashes of United Airlines Flight 585 in 1991 and USAir Flight 427 in 1994, and the near loss-of-control of Eastwind Airlines Flight 517 in 1996, appears to have been eliminated through a redesigned rudder power control unit, the results of the ETEB echo our findings that failure modes still exist in the Boeing 737 rudder system. While we are very concerned that some ETEB recommendations will not be adopted - particularly an independent switch to stop the hydraulic flow to the rudder and a rudder position indicator in 737 cockpits - we are pleased that both the FAA and Boeing Aircraft Company agree that there is a need for a redesign to the rudder actuator system. However, before the Board can determine if this will satisfy the goal of our recommendations, we will need to evaluate in detail the proposed design. I hope this redesign and retrofit can be accomplished expeditiously so that the major recommendation of our accident report last year will be realized - a reliably redundant rudder system for Boeing 737s.

On 13 November 2000 the Quick Reference Handbook Procedures for the 737 was revised again, when the FAA issued AD 2000-22-02 which superceded AD 96-26-07 'To require revising the FAA-approved Airplane Flight Manual (AFM) procedure in the existing AD to simplify the instructions for correcting a jammed or restricted flight control condition'.

That Airworthiness Directive was prompted by an FAA determination that the procedure currently inserted in the Aircraft Flight Manual by the existing Airworthiness Directive was not defined adequately. The actions required by that Airworthiness Directive were intended to ensure that the flight crew was advised of the procedures necessary to address a condition involving a jammed or restricted rudder.

A year later - on 5 June 2001 - the NTSB adopted a 'final final' report on the 1991 crash of United Airlines Flight 585 in Colorado Springs that called rudder reversal most likely cause.

On 12 November 2002 the FAA issuedAD 2002-20-07 R1; this was a revision issued a month after original AD:

'Within six years install a new rudder control system that includes new components such as an aft torque tube, hydraulic actuators, and associated control rods, and additional wiring throughout the airplane to support failure annunciation of the rudder control system in the flight deck. The system also must incorporate two separate inputs, each with an override mechanism, to two separate servo valves on the main rudder PCU; and an input to the standby PCU that also will include an override mechanism'.

In the first issue of AD 2002-20-07 it was stated: 'Because of the existing design architecture, we issued AD 2000-22- 02 R1 to include a special non-normal operational 'Uncommanded Rudder' procedure, which provides necessary instructions to the flightcrew for control of the airplane during an uncommanded rudder hardover event. The revised rudder procedure included in AD 2000-22-02 R1 is implemented to provide the flight crew with a means to recover control of the airplane following certain failures of the rudder control system. However, such a procedure, which is unique to Model 737 series airplanes, adds to the workload of the flightcrew at a critical time when the flightcrew is attempting to recover from an uncommanded rudder movement or other system malfunction. While that procedure effectively addresses certain rudder system failures, we find that such a procedure will not be effective in preventing an accident, the rudder control failure occurs during take-off or landing.

For these reasons, we have determined that the need for a unique operational procedure and the inherent failure modes in the existing rudder control system, when considered together, present an unsafe condition. In light of this, we proposed to eliminate the unsafe condition by mandating incorporation of a newly designed rudder control system. The manufacturer is currently redesigning the rudder system to eliminate these rudder failure modes. The redesigned rudder control system will incorporate design features that will increase system redundancy, and will add an active fault monitoring system to detect and annunciate to the flightcrew single jams in the rudder control system. If a single

A pair of Britannia Airways' 737-204s, with G-BTYF closest to the camera, seen at Luton in their later scheme. *(author's collection)*

A causeway just off the end of the runway at Corfu International Airport 'Ioannis Kapodistrias' allows for some interesting pictures, such as this rear-on shot of a Smartwings 737 starting it's take-off roll, complete with shimmering heat haze and swirling spray from the jet exhaust. It was in to this type of airport, on this type of service, that the 737 excelled. *(author)*

failure or jam occurs in the linkage aft of the torque tube, the new rudder design will allow the flightcrew to control the airplane, using normal piloting skills, without operational procedures that are unique to this airplane model'.

But this statement was withdrawn a month after it was issued saying 'Retaining this procedure will ensure that the flight crew continues to be advised of the procedures necessary to address a condition involving a jammed or restricted rudder until the accomplishment of this new AD'.

Chapter Seven

The -400/800 Series

Launch customer for the series 400 was Piedmont Airlines, which placed an order for twenty-five aircraft plus thirty options in June 1986.

Changes from the -300 airframe were small; notably, a tail bumper fitted to avoid damage in the event of grounding the rear fuselage on rotation. The wing spar was also strengthened to reflect the aircraft's increased weight, and each wing was equipped with an extra spoiler. Internally, the avionics were the same as the -300, but in response to demands from the airlines, the cockpit was fitted with 'glass cockpit' EFIS (Electronic Flight Instrumentation System) as standard.

The prototype -400 (c/n 23886), again wearing the N73700 registration, was rolled out at Renton on 26 January 1988, and took to the air for the first time less than a month later on 19 February, again with Jim McRoberts at the controls. Two aircraft undertook a seven-month, 500-hour test and certification programme.

The -400 was not an enormous seller in the USA but proved very popular with European and Asian operators. A number of major European airlines, including British Airways and Koninklijke Luchtvaart Maatschappij N.V. - KLM Royal Dutch Airlines - snapped up the aircraft since the extra capacity perfectly met their needs during the late 1980s and early 1990s. In the Far East, Malaysian Airline System accounted for no less than thirty nine series 400s, the largest operator of the type in the region. However, it was the European charter airline sector that really welcomed the aircraft. The 170 seats in an all-economy layout provided excellent cost savings for the operators while its range was sufficient to serve the popular Mediterranean holiday resorts from northern Europe.

On the flight deck, the most obvious difference of the -300/400 series from the -100/200 series is the Electronic Flight Instrument System (EFIS) displays,

BOEING 737-400

Length:	119ft 7 in (36.45 metres)	**Fuel (max):**	6,295 US gals (23,829 lit)
Height:	36 ft 6 in (11.12 metres)	**Range:**	2784 miles (4481 km)
Span:	94 ft 9 in (28.88 metres)	**Cruise Speed:**	0.745 mach
Empty Weight:	76,200 lbs (34,564 kg)	**Pax Capacity:**	146-170
Max Weight:	150,000 lbs (68,040 kg)	**Powerplants:**	2 x CFM56-3C-1

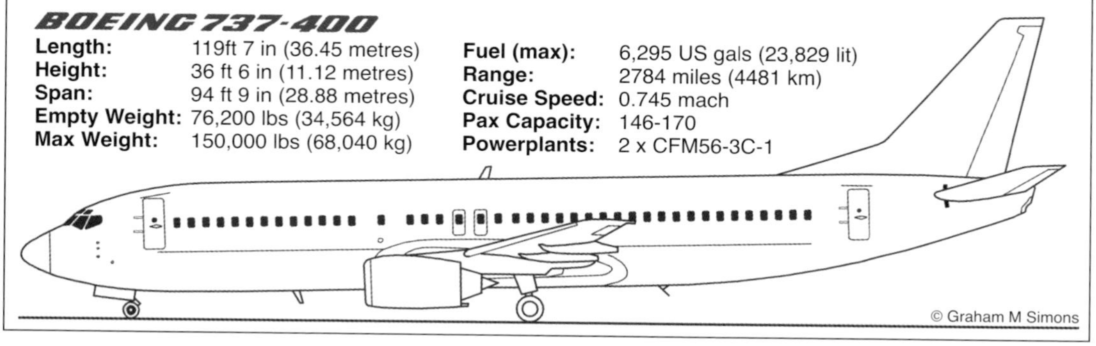

© Graham M Simons

Olympic Airways SX-BKE, a 737-484, has the cleaners board during a turnaround. *(author)*

although the very early -300 retained the electro-mechanical flight and engine instruments. Each pilot now had an Electronic Attitude Direction Indicator (EADI) which displayed the artificial horizon, speed tape, Localiser, Groundspeed, a Radar Altimeter and Mode Control Panel annunciations. Beneath the EADI was the Electronic Horizontal Situation Indicator (EHSI) which could display either navaid or route data superimposed with beacons, airfields, Flight Management Computer route - shown on the left-hand display - weather radar, terrain and Traffic Collision Avoidance System (TCAS) data.

On later -300 series aircraft, the digital Engine Information System display replaced twenty-one individual 'round dial' engine instruments. Post 9/11 modifications on some 737s included a locking flight deck door and Closed Circuit Television camera displays of the cabin with a monitor located between the Cockpit Display Units.

737-X

Boeing did not rest on its laurels but began work on a new version. The company reacted to increasing customer expectations, developments in new technologies and materials and to competition from Airbus Industrie when it approached over thirty airlines in 1991 for their views in defining the 737-X, as the Next Generation programme was then called. In June 1993 the Boeing Company board authorised the sale of the new aircraft that by now had assumed the name 737 Next Generation (NG). Although the basic design remained unchanged, the aircraft featured new

G-EZYC was another former Airlines of Europe aircraft, before going to Stelios Haji-Ioannou's low-cost carrier easyJet. Airlines of Europe was the overall 'airline' that incorporated Air Europe, Air Europa and Air Europe Italy. *(author's collection)*

Air Europe's 737-4S3 G-BPKA later went to Dan-Air and British Airways. *(Air Europe)*

wings, new engines, state-of-the-art cockpit technology and updated avionics. Initially, three variants were offered to the airlines: the -500X, which became the model 600 with up to 132 seats, the 737-300X, which became the -700 with up to 149 seats, and the -400X Stretch that became the -800 with up to 189 seats.

Although each of the original three variants was launched separately, they were effectively designed at the same time, and the flight test and certification programmes ran simultaneously.

The introduction of the latest advanced technology wings provided a cruise speed range of Mach 0.785 - 0.82 and a maximum altitude of 41,000 feet.

This aerodynamic efficiency helped to produce a thirty per cent fuel saving compared to older models. The wing was a pivotal element of the NG design as it afforded twenty-five per cent more wing area, and made use of double-slotted inboard flaps, a bigger aileron and new curved spoilers. Increased wing chord plus redesigned fuel tanks provided thirty per cent additional fuel capacity, and with the addition of winglet technology, performance has been even further enhanced. The optional winglets, originally developed for the BBJ, were eight feet tall and helped to reduce fuel consumption by nearly four per cent, increase payload range by as much as one hundred and twenty-five nautical miles and enhance take-off performance in hot-and-high locations. The benefits of the winglet technology are most pronounced on long-sector flights because they are most effective when the aircraft is established in the cruise. Short-haul operators almost invariably went for aircraft without winglets.

The pilots benefitted from an extensively modernised cockpit as well as a more capable aircraft. Considerable use was made of 777-style technology, with the latest in navigation and communications equipment, a full set of multi-function colour displays and new

BOEING 737-800

Length:	119ft 7 in (36.45 metres)
Height:	36 ft 6 in (11.12 metres)
Span:	94 ft 9 in (28.88 metres)
Empty Weight:	76,200 lbs (34,564 kg)
Max Weight:	150,000 lbs (68,040 kg)

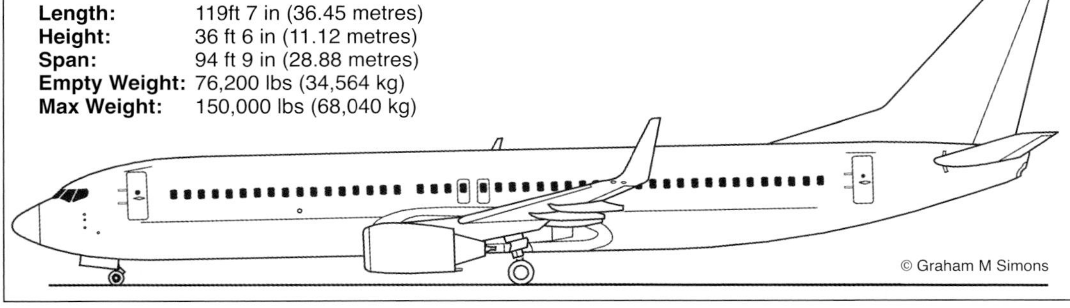

Image technology is making it harder and harder to decide if a picture is real or fake. This image of a jet2 737-800 is one such example. I lean towards fake, as the aircraft registration is not visible. *(jet2)*

Although high-density, the interior of jet2 aircraft are light, bright and airy. *(Hugh Jampton)*

avionics. However, Boeing decided to stick with a conventional flight control system for the 737NG rather than opt for the fly-by-wire route.

It took just a year to produce the first Next-Generation aircraft, which emerged from the Renton factory on 8 December 1996. Two months later, on 9 February 1997, Mike Hewitt and Ken Higgins took the prototype 737-700 N737X for its first flight, landing back at Boeing Field after an uneventful sortie. The first - 800, N737BX flew on 31 July 1997, and the first -600, N7376, made its first flight on 22 January 1998.

737-800

Just as the series 300 evolved into the series 700, so the series 400 morphed into the series 800. The 737-400X became the 737-800 and was significantly longer at 129 feet 6 inches and seating for up to 189 passengers. The project was launched on 5 September 1994, at the Farnborough Air Show with commitments for over forty machines. First flight was on 31 July 1997 and first delivery, to the German airline Hapag-Lloyd, was in April 1998.

Differences from the -700 included:

• Fuselage plugs of 9 feet 10 inches forward and 9 feet 4 inches aft of wing.

Additional pair of overwing exits added .
- Tailskid added to section 48.
- Environmental Control System riser ducts added.
- Re-gauged skins and stringers to strengthen wing and centre section.
- Strengthened main landing gear structure.
- 44.5 inch tyres, heavy-duty wheels and brakes.
- Engine thrust increased to 26,400 pounds.

737-800ERX

This was a heavier 184,000 pounds Maximum Take-Off-Weight, longer-range version of the -800 designed to meet the needs of the Multimission Maritime Aircraft. It was to have various components from the -900ER including its heavier gauge wing, nose and main gear and section 44 wing-body join section. It will also have some parts from the BBJ1. Unique features to the 800ERX would include strengthening to the empennage.

737-800SFP

The Short Field Performance improvement package was developed in 2005/6 to allow GOL airlines, a Brazilian low-cost airline based in Rio de Janeiro, to operate their 737-800s onto the 4,800 foot-long runway Santos Dumont airport. The modifications enabled weight increases of approx 10,000 pounds for landing and 3,750 pounds for take-off from short

Some airports, while not exactly marginal for 737 operations were, shall we say 'tight'? One such location was Skiathos Alexandros Papadiamantis Airport on the Hellenic island of the same name. At one end of the runway was a road, and then the harbour; at the other was a cliff and then the sea. There was no taxiway paralelling the runway, so every movement was 'back-tracked' and the airport apron was 'small' - so much so that during the early 2000s it often filled up, resulting in diversions to Thesssaloniki until space became available. Here excel airways 737-8Q8 G-EXAD negotiates its way up to a parking spot. *(author)*

Nova Airlines AB, operating as novair, is a Swedish airline headquartered in Stockholm, that operates on behalf of one of Sweden's largest travel agencies, Apollo, which is also the parent company. Here SE-DVO, a 737-85F, is seen at Samos Airport. *(author)*

runways. It included the following changes:

• Flight spoilers capable of sixty-degree deflection on touchdown by addition of increased stroke actuators. This compared to the previous 33/38 degrees and reduced stopping distances by improving braking capability.

• Slats were sealed for take-off to flap position 15 to allow the wing to generate more lift at lower rotation angles.

• Slats only travel to Full Extent when the Trailing Edge flaps were beyond twenty-five degrees. Autoslat function was available from flap one to twenty-five degrees.

• Flap load relief function was active from flap ten degrees or greater.

• Two-position tailskid that extended an extra five inches for landing protection. This allowed greater angles of attack to be safely flown thereby reducing Vref, the Landing Reference Speed at a point 50 feet above the landing threshold. It was

A long way from home! Sunwing Airlines offers scheduled and charter services from Canada and the United States to destinations within the United States, Mexico, the Caribbean, Central America, and South America. During the summer months, the company offers domestic services across Canada as well as services to European cities. The airline also leases its aircraft out, some of them roaming far afield, like 737-800 C-FWGH, seen landing at Corfu in Greece. *(author)*

enter air Sp. z o.o. is a charter airline with its head office in Warsaw, Poland, and main base at Warsaw Chopin and Katowice-Pyrzowice since its inaugural flight on 25 April 2010. It operates holiday and charter flights out of Polish and other European airports to popular Mediterranean holiday destinations. SP-ENN is a 737-8CX(W) and is seen landing at Corfu. *(author)*

not less than 1.3 times the stall speed in the normal landing. Use of this reduced the landing distance, and was monitored by a Supplemental Proximity Switch Electronic Unit SPSEU).

• Main gear camber (splay) reduced by one degree to increase uniformity of braking across all Main Landing Gear tyres.

• Reduction of engine idle-thrust delay time from five to two seconds to to shorten the landing roll.

• FMC & FCC software revisions.

The SFP package later became an option on all 737-800s - that were then known as 737-800SFP and standard on the 737-900ER. Some of the features were also fitted to the 600/700 series. The first SFP was delivered 31 June 2006.

For a short period following 9-11, the airliner market was very depressed and in

Lining up in the numbers for a take-off from Corfu is YR-BMB, a 737-85R(W) of Blue Air, a Romanian airline headquartered in Bucharest, with its hub at Henri Coandă International Airport. *(author)*

A stunning nose-on shot of an Air Europe 737. *(AE via author)*

shock, but by late 2002 the market was regaining confidence. New orders began to be placed by many of the world's airlines, although in the USA the market remained depressed, with many carriers deferring deliveries. Somewhat bucking the trend, the low-cost sector experienced considerable growth, especially in Europe, with a number of operators securing lucrative deals with Boeing. The biggest of these was placed on 24 January 2002, by Ireland-based Ryanair, a firm order for 100 737-800s with an option on a further 50. This is an indication of the popularity of the larger capacity -800 model, which has seen the most orders to date and is indicative of the growth in short-haul passenger numbers.

MMA / P-8A 'Poseidon'

Jack Zerr, the Multi-mission Maritime Aircraft (MMA) programme manager, described the aircraft as 'A bit of JSTARS (Joint Surveillance Acquisition Radar System), a bit of AWACS and a bit of MC2A (Multirole Command and Control), but with the added ability to go and kill a submarine.'

The Poseidon first flew on 25 April 2009 and was operational by late 2013.

The aircraft was based on the 737-800 fuselage and the stronger 737-900 wing, with raked wing-tips that have anti-icing along all leading-edge slats. A weapons bay aft of the wing - effectively in the aft hold varied in size, depending on internal stores carriage needs. There were four weapons stations under the wings. The fuselage was strengthened for weapons employment and to permit Anti-Submarine Warfare profiles. All maximum weights were significantly higher than a standard -800.

To reduce development costs, Boeing attempted to minimise aerodynamic changes from the -800, particularly with the nose cone. It worked closely with Raytheon to make their APS-137 search radar fit within the outer mould line. Up to seven mission consoles and a rotary sonobouy launcher were fitted in the cabin. Like the Airborne Early Warning and Command machines, the MMA had 180kVA Integrated Drive Generators as standard. The MMA was also designed with an in-flight refuelling receptacle over

ZP801, the first Poseidon MRA Mk. 1 *Pride of Moray* seen on a test flight. *(MoD)*

the flight deck.

Many of the modifications were done during production to save time and cost at the conversion stage. Boeing built a third production line dedicated to the MMA to allow this to happen alongside the commercial 737 assemblies. After the aircraft were assembled at Renton they were then flown over the Boeing Field for mission system installation.

Northrop-Grumman provided the electro-optical/infrared sensor, the directional infrared countermeasures system and the electronic support measures system. Raytheon were providing an upgraded APS-137 maritime surveillance radar system and signals intelligence (SIGINT) solutions. Finally, Smiths Aerospace provide the flight management and the stores management systems. The flight management system provides an open architecture along with a growth path for upgrades. The stores management system permitted accommodation of current and future weaponry.

Following an unveiling ceremony in Seattle, the first submarine-hunting Poseidon MRA Mk.1 Maritime Patrol Aircraft (MPA) was flown to Naval Air Station Jacksonville in Florida in Ocober 2019 where RAF personnel were being trained to operate the aircraft.

On arrival Michelle Sanders, Delivery Team Leader, signed the paperwork to formally transfer the aircraft, named *Pride of Moray*, to UK ownership. At the handover ceremony she said: 'Seeing the first Poseidon MRA Mk.1 handed over to the Royal Air Force is an incredibly proud moment for all of the team at DE&S. Close, collaborative working with colleagues in Air Capability, the US Navy and industry has helped us deliver this very capable aircraft.'

The MoD is investing £3 billion in nine state-of-the-art Poseidons, which will enhance the UK's tracking of hostile maritime targets, protect the British continuous at-sea nuclear deterrent, and play a central role in NATO missions across the North Atlantic.

Air Chief Marshal Mike Wigston, Chief of the Air Staff, said: 'Poseidon is a game-changing maritime patrol aircraft, able to detect, track and if necessary destroy the most advanced submarines in the world today. With Poseidon MRA1, I am delighted and very proud that the Royal Air Force will once again have a maritime patrol force working alongside the Royal Navy, securing our seas to protect our nation.'

First Sea Lord, Admiral Tony Radakin, said: 'Poseidon marks a superb upgrade in the UK's ability to conduct anti-submarine operations. This will give the UK the ability to conduct long-range patrols and integrate seamlessly with our NATO allies to provide a world-leading capability. This will maintain operational freedom for our submarines, and apply pressure to those of our potential foes. I look forward to working with the RAF and our international partners on this superb capability.'

The Poseidon MRA Mk.1 is designed to carry out extended surveillance missions at both high and low altitudes. The aircraft is equipped with cutting-edge sensors, uses high-resolution area mapping to find both surface and sub-surface threats.

The aircraft could carry up to 129 sonobuoys, small detection devices which are dropped from the aircraft into the sea to search for enemy submarines. The systems survey the battlespace under the surface of the sea and relay acoustic information via radio transmitter back to the aircraft.

The aircraft, armed with Harpoon anti-surface ship missiles and Mk 54 torpedoes was capable of attacking both surface and sub-surface targets.

The first aircraft was to arrive in Scotland in early 2020, with the fleet based at RAF Lossiemouth in Moray. All nine aircraft were to be delivered by November 2021.

The aircraft will be flown initially by 120 Squadron which was originally stood up on 1 January 1918 and was the leading anti-submarine warfare squadron in World War Two. 201 Squadron would join the programme later.

SIGINT

Boeing announced on 24 January 2006, the Signals Intelligence (SIGINT) variant, based on the MMA airframe for airborne intelligence, surveillance and reconnaissance, and also for advanced network-centric communications.

The MMA's sonobuoy system and anti-submarine warfare rotary launchers would be removed. The aft weapons bay would be sealed, and a small 'canoe' bulge would be added underneath to house a series of rotating SIGINT/ELINT antennas. The SIGINT aircraft would also have additional data links, embedded antennae, electronics and network-centric collaborative targeting capabilities to locate and identify signals, emitters and electronic attack packages to attack enemy devices and networks. There may also be AESA radars for their jamming, surveillance, and potential net-centric communications value.

The programme was cancelled in February 2010 and the US Navy fleet of EP-3E ARIES II were to be replaced with the MQ-4C Broad Area Maritime Surveillance (BAMS) unmanned aircraft and the MQ-8B Fire Scout unmanned helicopter by 2020.

Chapter Eight

The -500/600 Series

Both of the new-generation 737s so far produced were larger than the previous two models, but Boeing soon became very aware that many airlines required an aircraft closer to the size of the older Model 200 so as to meet their market requirements better. As a result, Boeing launched the 737-500 - a variant originally designated the -1000 - with a capacity of between 108 and 132 passengers. The aircraft was just one foot seven inches longer than the older -200 but otherwise was essentially the same airframe as the -300 and -400. Boeing test pilot Jim McRoberts was once more in command for the first flight of the prototype, on 30 June 1989. The obligatory N73700 registration was carried by the prototype, construction number 24178, but only this aircraft was used in the test and certification programme, which lasted seven months.

Launch customers for the -500 series were Braathens of Norway with twenty five aircraft and Southwest Airlines with twenty machines and it fell to Southwest to be the first to receive the type in February 1990. Despite the airlines apparently clamouring for an aircraft of this size, orders were somewhat disappointing when compared to the older -200, and only 389 series 500s were built. Sales were strong in the European market but elsewhere, even in the USA, take-up was light. This was partly due to the arrival in the mid-1990s of the Next Generation 737 programme, which saw the series 500 superseded by the series 600.

With the likelihood of a number of previously owned -300 and -400 models becoming available following the introduction of the 737NG, Boeing launched a 737-300/-400SF Special

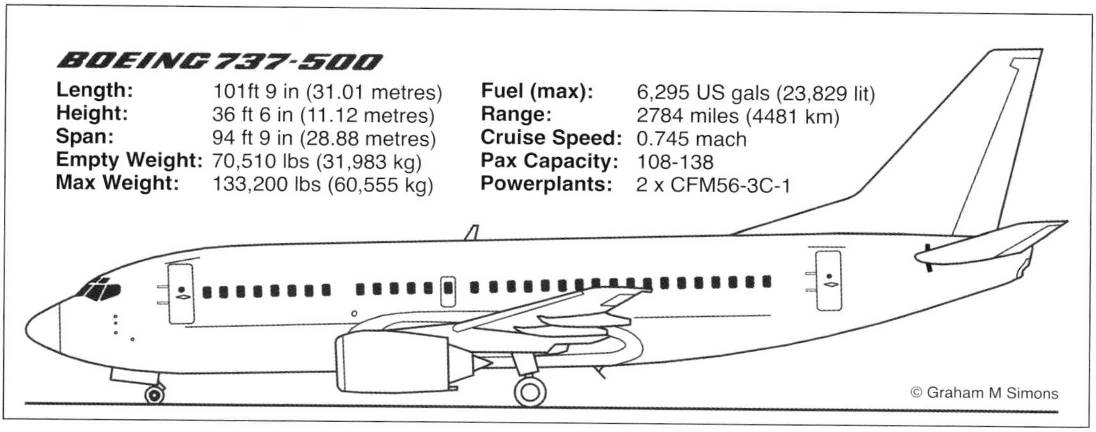
BOEING 737-500
Length: 101ft 9 in (31.01 metres)
Height: 36 ft 6 in (11.12 metres)
Span: 94 ft 9 in (28.88 metres)
Empty Weight: 70,510 lbs (31,983 kg)
Max Weight: 133,200 lbs (60,555 kg)
Fuel (max): 6,295 US gals (23,829 lit)
Range: 2784 miles (4481 km)
Cruise Speed: 0.745 mach
Pax Capacity: 108-138
Powerplants: 2 x CFM56-3C-1
© Graham M Simons

Southwest Airlines N511SW - a 737-5H4 - comes in to land. Southwest Airlines Co. is a major American airline headquartered in Dallas, Texas, and is the world's largest low-cost carrier. The airline was established on 15 March 1967 by Herb Kelleher as Air Southwest Co. and adopted its current name, Southwest Airlines Co., in 1971, when it began operating as an intrastate airline wholly within the state of Texas, first flying between Dallas, Houston and San Antonio. The airline has more than 60,000 employees as of the end of 2019 and operates about 4,000 departures a day during peak travel season. *(Matt Black Collection)*

Freighter programme with a capacity of eight or nine standard cargo pallets and a range of over 2,500 miles. The addition of the cargo door in the port side of the front fuselage and the fitting of a roller floor comprised most of the work in the conversion.

There is little doubt that the new generation series was a major boost for the Boeing Company and in February 1990 737 orders totalled almost 2,800, with deliveries exceeding the 1,832 total achieved by the 727. At this point, the 737 became the most popular jet airliner in the world, a mantle it maintained for many years. The last of the new generation aircraft was a 737-455 (OK-FGS, c/n 28478) for CSA Czech Airlines, rolled out at Renton on 9 December 1999, and delivered to the European carrier on 25 February 2000. With the arrival of the 737NG, the new generation -300/-400 and -500 aircraft became known as 737 classics.

In August 1992, Phillip Condit was named as president of the Boeing Company; then in December 1993, Ron Woodard was named as president of the Commercial Airplane Group, succeeding Dean Thornton, who retired in early 1994. Supersalesman Woodard set his sights on increasing market share.

In the two years of 1993 and 1994, Boeing slashed employment by 29,000. Factory costs had been dramatically reduced, something that reflected in the price of airliners, and increased competitiveness.

Boeing were still locked into an outmoded production system dating back to the days of World War Two. Getting from the initial order by the customer to the delivered article was a costly, complex system. In 1994, the Company began an initiative, scheduled for completion in 1998, aimed at simplifying and streamlining the complex process of engineering each airliner order to exact customer specifications.

Furthermore, the new system would simplify the processes used to schedule and order parts - including items that were outsourced - and manage the entire inventory. The improved method,

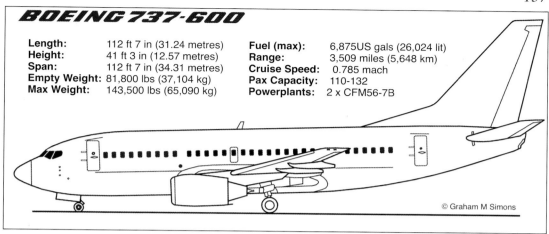

BOEING 737-600

Length:	112 ft 7 in (31.24 metres)	**Fuel (max):**	6,875US gals (26,024 lit)
Height:	41 ft 3 in (12.57 metres)	**Range:**	3,509 miles (5,648 km)
Span:	112 ft 7 in (34.31 metres)	**Cruise Speed:**	0.785 mach
Empty Weight:	81,800 lbs (37,104 kg)	**Pax Capacity:**	110-132
Max Weight:	143,500 lbs (65,090 kg)	**Powerplants:**	2 x CFM56-7B

combined with a switch to more efficient, 'lean' manufacturing techniques, was expected to significantly reduce both the time and cost required to produce airliners.

Growing by stages, the old system of effectivity callouts on the engineering drawings of the Stratoliner days had evolved into about four hundred separate computer systems used in 1994.

The new system, designed to replace all the others, had a tongue-tying acronym - DCAC/MRM - pronounced DeeKak-EmRam, for Define and Control Airplane Configuration/Manufacturing Resource Management. In simple terms, the system consisted of three Tailored Business Streams known by the acronym TBS.

TBS 1 would include all components

Ukraine International Airlines PJSC, often shortened to UIA, is the flag carrier and the largest airline of Ukraine, with its head office in Kiev and its main hub at Kiev's Boryspil International Airport. It operates domestic and international passenger flights and cargo services to Europe, the Middle East, the United States, Canada, and Asia. UR-GBC, a 737-5L9(W), originally built for Maersk Air is seen taxiing by the camera. *(Matt Black collection)*

LN-RCU, a 737-683 of Scandanavian Airlines - or to give it its more formal title SAS AB, or Scandinavian Airlines System Aktiebolag, seen about to land at London Gatwick. The conglomerate was founded in 1951 as a merger between the three Scandinavian flag carriers Aerotransport (ABA-Sweden), Det Danske Luftfartselskab (DDL- Denmark), and Det Norske Luftfartselskap (DNL - Norway), after the three had been cooperating on international routes since 1946. *(Kirk Smeeton Collection)*

common to all versions of a major model, such as the 737, which would include wing and body structure, empennage, landing gear, and many other parts.

TBS 2 would include all components common to a sub-model, such as the 737-500, -600 or whatever, which had already been designed, manufactured, and certified for a previous customer. TBS 3 would include customer-driven, newly defined components not previously designed, manufactured, or certified.

According to James 'Jim' Jamieson, Executive Vice President, the data from DCAC/MRM would then drive the system from the 'as defined' configuration entirely through the manufacturing cycle, identifying all parts and processes, and end with an 'as built' configuration.

EgyptAir - formerly Misr Airwork and Misrair - is the state-owned flag carrier of Egypt. It has been around since 1932, and the airline is currently headquartered at Cairo International Airport, its main hub, operating scheduled passenger and freight services to more than seventy-five destinations in the Middle East, Europe, Africa, Asia, and the Americas. Egyptair is a member of Star Alliance and the Arab Air Carriers Organization. Here SU-GBH, a 737-500 is seen returning from another service.

Chapter Nine

The -900 Series

The fourth and final variant in the NG family was the -900, which was launched on 10 November 1997, with an initial order for ten aircraft from Alaska Airlines. The -900 featured an increase in length to 138 feet 2 inches, giving room for up to 189 seats in an all-economy layout or typically 177 in a two-class layout. However, emergency egress regulations prevenedt the carriage of more than 189 passengers with the existing number of exit doors. Boeing has since proposed a 737-900X, which boasted increased range and a capacity of up to 220 passengers if fitted with an additional pair of type 1 doors aft of the wing.

Although each of the original three variants was launched separately, they were effectively designed at the same time and the flight test and certification programmes ran simultaneously.

Alaska Airlines became the launch customer for the 737-900 in 2001, when Iditarod Trail sled-dog race champion Doug Swingley and his team escorted the first aircraft out of the factory for its world premiere rollout celebration in Renton, Washington State.

The Boeing 737-900 featured 178 leather Recaro seats, seatback power outlets for charging laptops and smartphones, Premium Class, Boeing's award-winning Sky Interior, inflight internet service, and Alaska Beyond Entertainment. As the airline said not long after they put the series 900 into service: 'We're also adding Boeing's innovative Space Bins to these aircraft. The larger overhead bins hold forty-eight percent more bags, allowing customers to load bags with less hassle'.

KLM: The Boeing 737-900 is the 3rd Boeing 737 model that we've added to our fleet, the 1st one of this type joined in 2001. Currently, we've got five of them. This -900 might just fly you to beautiful

BOEING 737-900

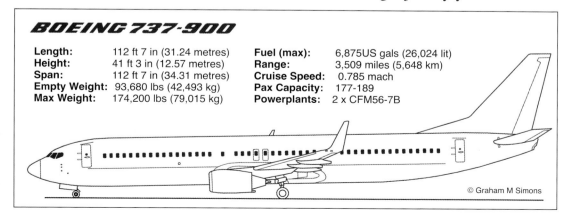

Length:	112 ft 7 in (31.24 metres)	**Fuel (max):**	6,875US gals (26,024 lit)
Height:	41 ft 3 in (12.57 metres)	**Range:**	3,509 miles (5,648 km)
Span:	112 ft 7 in (34.31 metres)	**Cruise Speed:**	0.785 mach
Empty Weight:	93,680 lbs (42,493 kg)	**Pax Capacity:**	177-189
Max Weight:	174,200 lbs (79,015 kg)	**Powerplants:**	2 x CFM56-7B

© Graham M Simons

Above: 737-990ER (W) N434AS of Alaska Airlines. *(Alaska)*

Right: the serried ranks of economy seating await the passengers. *(Matt Black collection)*

Tel Aviv, Stockholm or Bucharest, although these destinations may differ per season. The Boeing 737-900's, just like the -700's and -800's, are all named after birds: *Plover, Crested Coot, Nightingale, Buzzard,* and *Sea Tern*.

United Airlines operated three slightly different interior configurations of the series 900. The number of 'United First®' class passengers were all the same, but the difference in United Economy Plus® and United Economy® varied slightly.

Boeing began work on the 737-900 in April 1997 which was stretched to compete with the 185/220 seat, Airbus A321. It featured a seven foot ten-inch fuselage extension giving it an overall length of 138 ft 3 inches actually making it longer than the 707-120. The 900 has nine oper cent more cabin floor space and eighteen per cent more cargo space than the -800; however Boeing opted to use the same NG emergency exit layout, with four main exit doors and four overwing exits, thereby still restricting the maximum passenger load to 189. Due to slow sales, it was succeeded by the 737-900ER.

737-900F

Boeing revealed the 737-900F study group in September 2003. The project was believed to be aimed at FedEx who was looking to replace their 727 fleet. It would use the side cargo door from the 700QC and be capable of taking eleven standard 2.24 x 3.18m pallets, three more than the -700QC and only three less than the 757. The -900 also has a hold volume of 51.7cu.m. This project was later shelved.

737-900ER

The -900ER (formerly known as the -900X) had the same length fuselage as the -900. Seating for up to 215 passengers has been achieved by adding a pair of Type II doors aft of the wing for passenger evacuation regulations and installing a new flattened aft pressure bulkhead which would add an extra fuselage frame (approx one row of seats) of cabin space. The flat bulkhead became standard on all 737s from 2006 and the Type II door also became standard on all series 900s although operators could choose to have it deactivated.

Range was increased to 3,200 nautical miles with the addition of two 1,970 litre auxiliary fuel tanks, or 2,800 nautical miles without and optional winglets. The 900ER had reinforced landing gear legs, wing-box and keel beam structure to handle the increased Maximum Take-Off-Weight of 187,700 pounds. Take-off and landing speeds, and therefore field length was reduced by the short field performance improvement package originally developed for the 737-800, and was standard on all 737-900ERs. Maximum Landing Weight of 157,500 pounds, making it similar in weight to the 727-200; the brakes were upgraded as a consequence.

Production started in 2006 followed by a two aircraft, seven month, flight test programme starting 1 September 2006. FAA certification was gained on 26 April 2007 with the first aircraft delivered to Lion Air the following day.

Evolution of Winglets.

The most noticeable feature to appear on 737s since 2000 were winglets. These are wing tip extensions which reduce lift induced drag and provide some extra lift. They have been credited to Dr Louis Gratzer formerly Chief of Aerodynamics at Boeing and later with Aviation Partners Boeing (APB).

The original winglet design was by NASA Langley aeronautical engineer Richard Whitcomb during the 1973 oil crisis. They were first flown on a 737-800 in June 1998 as a testbed for use on the BBJ. They soon became available as a standard production line option for all NGs with the exception of the -600 series. They are also available as a retrofit from APB. Each was around eight feet two inches tall, and about four feet wide at the base, narrowing to approximately two feet at the tip and added almost five feet to the total wingspan. The winglet for the Classic is slightly shorter at seven feet tall. Most 737NGs had winglets, and all MAX's were to be built with winglets.

Split Scimitar Winglets (SSWs) were offered by APB for the 737-800 and 737-900ER and came into service in early 2014. They were also offered as a retrofit to existing winglet aircraft.

A set of SSWs weighed 294 pounds per aircraft but gave fuel savings of 1.6 per cent on sectors of 1000 nautical miles rising to 2.2 per cent on sectors of 3000 nautical miles which equated to an extra sixty-five nautical miles of range. The modification required a trailing edge wedge, strengthened stringers and ballast weight but no changes to any avionics or the FMC.

The first United Airlines Boeing 737-800 fitted with new split scimitar winglets from Aviation Partners Boeing takes off on its maiden test flight on 16 July 2013. The aircraft flew from Paine Field in Everett, Washington According to United Airlines, the new winglet design improved on the existing blended winglets then fitted to the carrier's Boeing 737NG fleet. United was to serve as the launch customer for the new split scimitar winglet when it made a firm commitment with Aviation Partners Boeing to retrofit its Boeing 737-800 fleet and later announced it would also retrofit its Boeing 737-900ER fleet.

United's programme to retrofit its Boeing 737-800s and 737-900ERs with

Above: G-TAWW of TUI in what became known as 'the wave' colour scheme is seen on the turn-around at Corfu International Airport. The airliner is the former D-ATUQ, a 737-8K5(W) of TUIfly.

Right: a close-up of the split scimitar winglet. *(both author)*

split scimitar winglets consisted of replacing each existing blended winglet aluminium winglet tip cap with a new aerodynamically shaped 'Scimitar' winglet tip cap and adding a new Scimitar-tipped ventral strake 'We are always looking for opportunities to reduce fuel expense by improving the efficiency of our fleet. The Next-Generation 737 Split Scimitar Winglet will provide a natural hedge against rising fuel prices while simultaneously reducing carbon emissions' said Ron Baur, vice president of the fleet for United Airlines. The new split scimitar winglets with which United retrofitted to its Boeing 737-800 and 737-900ER fleets look similar to the winglets which featured on the 737 MAX family. United estimated each set of split scimitar winglets would reduce the fuel burn of any Boeing 737NG on which they are installed. With the split scimitar winglets are installed, United expected the winglet technologies installed on its 737NG, 757, and 767-300ER fleets to save it more than $200 million per year in jet fuel costs.

Chapter Ten

Hirings, Firings and Take-Overs

On 15 December 1996 Boeing Co - then the world's largest commercial aircraft maker - announced that that it would buy its long-time rival McDonnell Douglas Corporation for $13.3 billion in stock, so creating the world's largest integrated aerospace company.

The move left Airbus Industrie, the increasingly successful European consortium, battling head to head with only a single major competitor for the lion's share of the world's commercial airliner business.

It also made Boeing a more potent competitor against Lockheed Martin Corp. and others in the military sector. McDonnell Douglas had long been an important supplier of combat aircraft, not just to the USA, but to many allied nations, from Britain and Italy to Malaysia and Japan.

Boeing, with overflowing order books intended to use surplus factory capacity of the struggling McDonnell to make itself a more efficient producer. The companies would have a combined order backlog of about $100 billion.

Philip Condit, president and chief executive of Boeing, called the deal an historic moment in aviation and aerospace. He was appointed chairman and Chief Executive Officer of the combined company. Harry Stonecipher, chief executive of McDonnell Douglas, was to be president and chief operating officer.

Harry C Stonecipher was a former president and chief executive officer of American aerospace company McDonnell Douglas.. He began his career at General Motors' Allison Division, where he worked as a lab technician and was influenced by Jack Welch. He moved to General Electric's Large Engine Division in 1960 and began to move up the ranks. He became a vice president at GE in 1979, then a division head in 1984. In 1987 he left for Sundstrand, where he became president and Chief Executive Officer in 1989.

In September 1994, Stonecipher was elected president and Chief Executive Officer of McDonnell Douglas, holding this post until its merger with Boeing. During this period he became much more of a public figure and even began hosting the company's quarterly video report. He remained on the board following the successful completion of that transaction, serving as president and Chief Operating Officer of the merged entity.

Stonecipher orchestrated the merger between McDonnell Douglas and Boeing and was widely credited with the seeming resurgence of Boeing after government procurement scandals. However, his tenure also included major decisions to change Boeing's design and sourcing process for the new 787 'Dreamliner' airliner. These decisions later proved to be organisationally and financially disastrous for Boeing.

Stonecipher came out of retirement to lead Boeing, following the resignation of Chairman and Chief Executive Officer Phil Condit in December 2003 over several scandals. These scandals surrounded

Left: Philip Murray Condit (*b*. 2 August 1941)

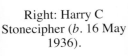

Right: Harry C Stonecipher (*b*. 16 May 1936).

allegations of documents stolen from competitors and the hiring of a government procurement officer who at the time, was involved in the United States Air Force's KC-767 contract. Stonecipher assumed the titles of President and Chief Executive Officer, which was not considered an interim appointment as there was no search initiated for a new chief executive, while Lewis Platt became non-executive chairman of the board.

Philip Murray Condit was an American engineer and businessman who was Chair and Chief Executive Officer of the Boeing company from 1996 to 2003. Condit joined the Boeing company in 1965 as an aerodynamics engineer and worked on the Supersonic Transport programme. The same year he was awarded a patent for a flexible wing design called a 'sailwing'. In 1968, he became a lead engineer on the Boeing 747 high-speed configuration. He advanced into management within a year, then became manager of Boeing 727 marketing in 1973.

In 1974, he entered the Sloan Fellows programme at the MIT Sloan School of Management, where he completed his Master's degree in management a year later. He returned to Boeing as manager of new-programme planning. He then advanced to director of programme management for the 707/727/737 division in 1976, and 757 chief program engineer in 1978, then director of 757 engineering in 1981.

Condit later became vice president and general manager of the 757 division in 1983; vice president of the Renton division the same year, and vice president of sales and marketing for Boeing Commercial Aircraft Company (BCAC) in 1984. In 1986, he was named as Executive Vice President and general manager of BCAC, then executive VP and general manager of the 777 division.

But back to 1996. It was said by corporate analysts that the combined company would be the 'largest, strongest, broadest, most admired aerospace corporation in the world and by far the largest US exporter,' said John McDonnell, chairman of McDonnell Douglas. The companies said they anticipated no plant closures or layoffs because of the deal.

Company executives revealed that discussions on a deal began in 1993. The takeover was subject to approval by regulatory agencies, who would be looking for possible antitrust problems. But Philip Condit said he did not expect the deal to have any problems in obtaining antitrust approval. '...We looked very carefully at all of our programs. We believe that dramatically they are complementary. There are some overlaps but they're very, very minor'.

Harry Stonecipher went on record saying 'In the military and space areas, we're absolutely complementary. There's plenty of competition in this arena'.

The buy-out deal saw Boeing giving McDonnell Douglas shareholders 0.65 Boeing shares for each McDonnell Douglas share. That was equal to $62.89 per share - a huge premium over McDonnell Douglas' closing price of $52.

The Group would continue to be known as Boeing; McDonnell Douglas retained its name and was to operate as a major division. Two-thirds of the board members came from Boeing, which retained its Seattle, WA headquarters.

The combined company had an estimated 1997 revenue in excess of $48 billion, with about $28 billion from Boeing and $20 billion from McDonnell Douglas. It would employ around 200,000 people, including those brought into Boeing by its then-recent purchase of Rockwell International Corporation's defence and space division.

For about eighty years, Boeing Aircraft functioned as an association of engineers. Its executives held patents, designed wings, spoke the language of aviation engineering and safety as a mother tongue. Finance wasn't a primary language. Even Boeing's bean counters didn't act the part. As late as the mid-'90s, the company's chief financial officer had minimal contact with Wall Street and answered colleagues' requests for even primary financial data with a simple 'Tell them not to worry.'

Richard Raymond Albrecht.

The 'Reverse Take-Over'

There then started a convoluted story that demonstrates a transition of power from a primarily engineering structure to one of accountancy, number-driven powerbase that saw corners cut, and the previous extremely high safety methodology compromised.

The take-over started to be called Boeing's 'reverse take-over' of McDonnell Douglas - an event so-called because it was McDonnell executives who perversely ended up in charge of the combined entity, and it was McDonnell's culture that became ascendant. 'McDonnell Douglas bought Boeing with Boeing's money,' was the joke that went around Seattle. Condit may have still been in charge, but Stonecipher was cutting a Dick Cheney-like figure, blasting the company's engineers as 'arrogant' and spouting Harry Trumanisms such as 'I don't give 'em hell; I just tell the truth, and they think it's hell' when they shot back that he was the problem.

It was now time for the hard part - performing. Even though the transaction was a buyout of McDonnell Douglas by Boeing, both companies wanted desperately for the marriage to become a true merger - from the top all the way to the factory floor.

McDonnell's stock price had risen fourfold under Stonecipher as he went on a cost-cutting rampage, but many analysts feared that this came at the cost of the company's future competitiveness. 'There was a little surprise that a guy running a failing company ended up with so much power,' the former Boeing executive vice president Dick Albrecht was reported to have said at the time. Post-merger, Stonecipher brought his butchers knives to Seattle. 'A passion for

737 fuselages on custom-built railcars are pushed into the Boeing plant by a BNSF locomotive. *(Simon Peters collection)*

affordability' became one of the company's new, unloved slogans, as did 'Less family, more team'.

For many reasons - some said preoccupation of the top managers with the merger - earnings took a nosedive, and Boeing reported its first net loss in fifty years, recording $178 million in red ink in 1997. The performance was so bad that Boeing was the 'dog of the Dow' in 1998, with its stock falling more than thirty-three per cent. It seemed that Boeing had lost sight of why it was in business.

The Company, claiming that failing to do a good job in the management of the ambitious ramp-up in production of commercial aircraft was the main problem, promised to do better in 1998, pointing out that the backlog was the highest ever at $121.6 billion.

The 'new' Boeing had embarked on a nearly impossible programme aimed at more than doubling the number of aircraft coming off the production lines over a period of eighteen months. Many 737s were rolling out unfinished, sporting a blizzard of 'pickup' tags, showing that there were items still to be fitted or completed. These 'out-of-sequence' airliners had to be finished on the flight line, incurring inordinate costs of moving people and parts out of the final assembly building.

The Company was forced to take drastic measures - shutting down both the 737 and 747 lines for a month to bring the work back into sequence - the first shutdown since the end of the B-29 programme.

The cause of the problem was much more complex than simply the ramp-up decision. Between 1989 and 1995, employment had been reduced from 106,670 to 71,834 in Washington State, which included the 7,000 veteran employees who had taken early retirement in 1995 - a wealth of knowledge and experience that the Company could ill afford to lose in one grand swoop. Then, between 1995 and 1998, employment was increased by 28,000. Many of these new people required extensive training, further stretching the Company's thin line of expertise.

Further adding to the pressure was Boeing's decision in 1996 to focus on market share. With Ronald B 'Ron' Woodard, President of the Boeing Commercial Aircraft Group in the driver's seat, and financial discipline all but abandoned after the retirement of Harold Haynes, an unchecked drive to defeat Airbus Industrie escalated to the point where 737s were being sold at breakeven prices - in some cases even below cost. In 1997, fixed-price contracts were negotiated with American, Continental, and Delta wherein Boeing would be their sole supplier for the next twenty years. Woodard believed that the Company could overhaul its antiquated production and inventory control methods in time to recover profits by reducing manufacturing costs by twenty-five per cent, while at the same time delivering a record number of airliners. He hoped to bury Airbus in an avalanche of orders, predicting confidently, 'Our goal is a sixty-seven per cent market share.'

On 25 August 1998, British Airways announced it would purchase Airbus Industrie narrow-body aircraft instead of Next-Generation 737 airliners, a complete shock for Boeing, since the airline had been an exclusive Boeing customer. By this time, top management at Boeing had recognized they were pursuing a bankrupt course and had decided to draw a line in the sand - refusing to continue the price war.

The 'old hands' were less than kind to Boeing management. At the spring banquet of the Gold Card Chapter of the Boeing Management Association (BMA) in 1998, the guest speaker was Jim Dagnon, newly hired with the title of Senior Vice President, People. Dagnon had been active in leading the successful merger of Burlington Northern and Santa Fe railroads, and his experience as a leader in merging corporate cultures was felt to be invaluable. When Dagnon finished his speech and invited questions, he was verbally attacked from all sides.

'What's all this crap about people organizations,' one irate retiree wanted to know. 'Why don't you just get down to the brass tacks of building airplanes?'

Another veteran spoke up. 'What in hell is the matter down at Boeing? We never missed a delivery date in twenty years.'

Dagnon, taken aback by the intensity of the unexpected criticism, attempted to explain, but the audience didn't seem to be listening.

Contributing to the dismay of the old timers was the bewildering new titles for management people. In addition to 'People', there was 'Workforce Administration',

Left: Ronald B 'Ron' Woodard

Right: James 'Jim' Dagnon.

'Advertising and Corporate Identity,' and 'Socioeconomic Executive'... The list was almost endless. It even went beyond management. The age-old title of secretary was now 'Office Administrator', in deference to the newly elevated philosophy of 'personal esteem' as the most important virtue of the workforce.

There was no simple way to explain to the retirees what was happening at the Company. They represented a radically different Boeing - a company that was strongly engineering-oriented, where the men in suits and ties made the decisions, and the people in the factory went out and got the job done, no questions asked. Boeing was still in the process of reinventing itself to accommodate the vast cultural shifts in the USA of the previous decades. This 'new' Boeing was being called upon to assimilate two companies, each with a corporate culture of its own.

Jim Dagnon pointed out that Boeing now faced a much different marketplace. 'Just being a great engineering Company is no longer sufficient. Now we are a manufacturing Company facing the pressures from Airbus - a vigorous and competent competitor. Our priorities are how we control our cost, and how we control our inventory.'

Engineers were all too happy to share such views with executives, which made for plenty of awkward encounters in the still-smallish city that was Seattle in the '90s. It was, top brass felt, an undue amount of contact for executives of a modern, diversified corporation.

One of the most successful engineering cultures of all time was quickly giving way to the McDonnell mind-set. Another McDonnell executive had recently been elevated to the chief financial officer, something that was a further indication of who was controlling this company.

During the massive increase in sales of the late eighties, organizations expanded vertically, and new layers appeared all through the Company. *Boeing News,* a bellwether of Company fortunes, showed the trend, when in October 1997 the page count jumped from sixteen to twenty, and occasionally twenty-four, during 1998. Indeed, the 19 January 1998 issue revealed a complete overhaul of the newspaper. It was announced that more local news would be included, and more news about people. Thus, separate local editions were introduced for the employees in California, the Northwest States, the Rotorcraft operations, St. Louis, and Wichita. The new-look would include colour on all pages.

Paradoxically, in spite of promises to do better, Boeing's 737NG programme became

Typical of many an airport scene in the USA at the turn of the century - a pair of Delta airliners in different colours, 737-832(W) N3745B, and a 727. *(author)*

Delta 737 landing at Portland OR, with spoilers deployed. *(author)*

The interior of a Delta Airlines 737-900, looking rearwards from the First Class section. *(author)*

the Company's most financially troubled, at the same time gaining the most early sales of any project the Company had ever undertaken. In the spring of 1998, Boeing announced it would write-off up to $350 million in the first quarter. It was related to production and certification delays, parts shortages, and late delivery penalties to customers. It came on top of a similar $700 million write-off against the 737NG in 1997. Boeing had sold 866 Next-Generation 737s by the end of the first quarter, but according to Wall Street analysts, the company stood to lose $1 billion on the first 400 machines.

It was enough to drive the white-collar engineering union, which had historically functioned as a professional debating society, into acting more like organised labour. 'We weren't fighting against Boeing...' one union leader said of the forty-day strike that shut down production in 2000. 'We were fighting to save Boeing'.

Boeing began cleaning up its operations. In June they announced the end of production of the MD-11 for February 2000, completing orders on the books.

Sales of the MD-95, which was destined to be the last 'MD' model to go into production, had an auspicious start with an AirTran Airlines order for fifty, but soon slowed to a trickle. To improve its image, the MD-95 label was dropped in favour of a venerable old 700 series number - the Model 717. The designation had been around since the beginning of the commercial jet age, first assigned to the tanker version of the 367-80 prototype, later named the KC-135 by the Air Force, and still later offered to and turned down by United Airlines in favour of the Model 720 designation.

In May 1998, when Boeing revealed a plan to transfer some 737 work to the Douglas Long Beach plant and possibly opening a new final assembly line there, the

Seattle local of the International Association of Machinists vehemently objected, threatening to take the Company to court.

After intensive negotiations, the union agreed not to fight the Company's plans, in exchange for Boeing's promise to limit production there to just five airplanes per month. This agreement was the final hurdle Boeing needed to clear, and the Company finalized plans to implement a production line at Long Beach in the autumn.

Ten 'out-of-sequence' 737s were flown to Long Beach to be completed and readied for delivery, thus taking some early pressure off the congested Renton production lines. A csmall group of Douglas mechanics were trained by Boeing supervisors from Seattle to complete the installation of galleys, lavatories, and seats, and to finish up other interior work.

Nevertheless, with relentless Wall Street pressure for higher earnings, there seemed to be no doubt that some heads in the Commercial Airplane Group would roll.

Abruptly, on 2 September 1998, Ron Woodard, president of the Commercial Airplane Group, was asked to resign, as were members of his top team. In his place came Alan Mulally, only recently having taken charge of the Information, Space, and Defense Group.

Ron Woodard was fired with strong reluctance by both Phil Condit and Harry Stonecipher, since the drive for market share was fully endorsed by top management. The financial community demanded a sacrificial lamb and from a high level and so Woodard was thrown to the wolves.

Boeing awarded him a handsome separation package. He received $900,000 in cash, and a consulting fee of $19,000 a month through to November 2000, or about $450,000. He also received $43,000 to pay taxes on stock options he had recently exercised before he was fired and matching recent contributions he had made to local charities. The severance totalled about $1.35 million.

Mulally's first action after taking over was to call Bill Johnson, president of the IAM local in an attempt to cement Union/company relations and this contact demonstrated a genuine effort to improve relations with the Company's largest union. The four-year contract was scheduled to expire on 1 September 1999, and both the union and Boeing management was determined to avoid a strike. Bill Johnson, a veteran Boeing employee who joined the Company in 1965, was painfully aware of the cyclical nature of airliner production and recognized that job security would be one of

Left: Alan R Mulally (*b*. 4 August 1945), who replaced Ron Woodard as President of the Commercial Airplane Group.

Right: James A Bell (*b*. 4 June 1948), President, Executive Vice President and Chief Financial Officer of The Boeing Company. He retired, in 2012.

the strongest concerns of his union members. Work outsourced to Chinese factories was a major issue in the 1995 strike. To demonstrate the importance of this practice, Boeing sponsored a joint trip to China in the autumn of 1998 to understand first hand why outsourcing was so important to Boeing sales in the world's airliner market. Fifteen members of the union were on the team which visited all the aircraft factories in China.

Both sides began conciliatory talks early in 1999, leading to contract negotiations in August. The negotiated result over the 737 Long Beach production controversy gave cause for optimism.

It was announced that Harry Stonecipher would take over as acting financial officer, as Boyd Givan was forced out, retiring early on 1 September 1998. In November, Deborah Hopkins was introduced as the new chief financial officer for Boeing, the highest-ranked woman ever employed by the Company. She came with impeccable credentials, most recently as vice president of General Motors European operations.

Hopkins had received a bachelor's degree in business from Walsh College and attended the Wharton School of Business. Boeing Chairman Philip Condit praised Hopkins for her track record at GM, pointing out her leadership in adopting strategies for lean manufacturing processes pioneered by the Japanese. She came at a base salary of $450,000 and a guaranteed performance incentive of $360,000 for 1999, which could be significantly higher, depending on business results, as well as a signing bonus of $750,000.

Reorganization of the company continued apace. Boeing had been searching for someone with a high profile in communications, and in December 1998, when 62-year old Larry Bishop announced his early retirement, Boeing quickly hired Judith Muhlberg, a twenty-two year veteran of the Ford Motor Company. She brought an impressive curricula vitae to the job, including two years in the White House during the Ford administration. A graduate from the University of Wyoming in communications, Muhlberg was a Fulbright scholar and held a Juris Doctorate from the Detroit College of Law.

As Vice President, Communications, Muhlberg was given overall responsibility for the Company's public relations, executive and employee communications, and advertising. In September, Boeing announced the appointment of two new vice presidents, reporting to Muhlberg: Anne Toulouse, as Vice President, Advertising and Corporate Identity, and Thomas J 'Tom' Downey as Vice President, Executive and

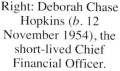

Left: Judith Muhlberg, who became Vice-President, Communications.

Right: Deborah Chase Hopkins (*b*. 12 November 1954), the short-lived Chief Financial Officer.

Internal Communications, so separating employee communications from the Company's external media and advertising efforts, which were expected to exert a greater global reach in concert with McCann Erickson, its new advertising agency.

Boeing employees were quick to note that Toulouse and Downey came from the old McDonnell Douglas organizations. Toulouse, after receiving her bachelors degree in science from Florida State University in 1980, and serving as a media specialist for the USAF, joined McDonnell Douglas in 1989 and rose to Director of Communications at Huntington Beach. Tom Downey, was a graduate of St. Louis University with a degree in English, joined McDonnell Douglas in 1986 as an associate writer on the communications staff. He took a one year leave of absence in 1990 as a Brookings Institute Congressional Fellow. Previously he was general manager of Communications and Community Relations for the Military Aircraft and Missile Systems Group at St. Louis.

In the wake of the management shakeup of Boeing's Commercial Airplane Group, the Company began rethinking its plans to open a new 737 line in Long Beach, and notified factory workers there that startup was being postponed until early 1999, possibly in February.

When Boeing executives visited the Long Beach plant, they were appalled to find how much work and investment would be required to bring it up to Boeing standards, and on 11 December 1998, Boeing announced that all final assembly work for the 737 would remain at Renton because of '...excessive incremental costs for the Long Beach operation'.

Boeing had been intent on overhauling its financial structure ever since the retirement of Harold Haynes in 1997. When Debbie Hopkins was brought in as Chief Financial Officer, a move attributed to Harry Stonecipher, she hit the ground running. In short order, she revamped Boeing's financial management and instituted tough new financial goals and a performance-oriented culture. From the beginning, she had her eye set directly on CEO Philip Condit's job but soon found that was not to be.

In April 1999, she abruptly resigned her Boeing post to accept the CFO job at Lucent Technologies, after a tenure of just fifteen months. That company did not do well and she was fired after one year. A week later there was an even bigger surprise when Boeing announced that Michael Sears, a man without a financial background, would be the new Chief Financial Officer. Sears had been head of the St. Louis operations of McDonnell Douglas, reporting directly to Harry Stonecipher. It is not hard to see why many thought that it was McDonnell Douglas that had taken over Boeing!

Under Stonecipher's tenure, the Air Force had lifted a twenty-month suspension of Boeing's Launching Systems Group, which had been involved in one of the scandals, allowing them to bid on Pentagon contracts again. He also oversaw the launch of the Boeing 787 Dreamliner to challenge Airbus. Although not fully evident at the time, the results of significant changes to Boeing's airliner programme design, sourcing and financing made during Stonecipher's and Condit's tenures would later prove disastrous. Shares of the company traded as high as $58.74 in 2005, up fifty-four per cent during his tenure.

Stonecipher submitted his resignation upon request of the Boeing Board of Directors on 6 March 2005, after an internal investigation revealed a consensual relationship with Boeing executive Debra Peabody. The probe found that Boeing business operations were unaffected, that Peabody's career and compensation were not influenced, and that there was no improper use of company expenses or property. Nonetheless, the board of directors decided that there would be 'zero tolerance on breaches of ethics'. His wife of fifty years, Joan Stonecipher, filed for divorce just days after news of his affair became public. Stonecipher was succeeded as president and CEO on an interim basis by Chief Financial Officer James A. Bell until James McNerney was hired on a full-time basis.

After a long and distinguished career, Stonecipher was forced to resign at Boeing following the disclosure of his longtime consensual affair with the fellow Boeing employee whom he later married.

A flight to disaster?

The event that sent Boeing off course was a flight that put the company on a path for disaster. It lifted off a few hours after sunrise in May 2001, but no one knew where the 737 was headed. The flight crew had prepared three flight plans: one to Denver, one to Dallas and one to Chicago. On board were company executives and representatives of the national, international and technical media. This flight - little more than a PR stunt to end the two-month contest for Boeing's new headquarters - would reveal the answer.

In the airliner's trailing vortices was greater Seattle, where the company's famed engineering culture had taken root; where the bulk of its 40,000-plus engineers lived and worked; indeed, where the jet itself was assembled. Boeing's leaders, CEO Philip Condit and President Harry Stonecipher, had decided it was time to put some distance between themselves and the people making the company's airliners. How much distance? Once the aircraft was airborne, Boeing announced it would be landing at Chicago's Midway International Airport.

On the tarmac, Condit stepped out of the jet, made a brief speech, then boarded a helicopter for an aerial tour of Boeing's new corporate home: the Morton Salt building, a skyscraper sitting just out of the Loop in downtown Chicago. Boeing's top management plus staff, roughly 500 people in all, would work here. They could see the boats plying the Chicago River and the trains rumbling over it. Condit, an opera lover, would have an easy walk to the Lyric Opera building. But the nearest Boeing commercial aeroplane assembly facility would be 1,700 miles away.

The isolation was deliberate. 'When the headquarters is located in proximity to a principal business - as ours was in Seattle - the corporate centre is inevitably drawn into day-to-day business operations,' Condit explained at the time. And that statement, more than anything else that had happened before or since, captured a cardinal truth about the aerospace giant. The present 737 MAX disaster can be traced back two decades - to the moment Boeing's leadership decided to divorce itself from the firm's own culture.

The move would let the devils triumvirate of Accountants, Public Relations Managers and 'Corporate Strategists' take the helm. If Andrew Carnegie's advice of 'Put all your eggs in one basket, and then watch that basket' - had guided Boeing before, then these decisions accomplished roughly the opposite. The company would put its eggs in three baskets: Military in St. Louis, Space in Long Beach and Passenger jets in Seattle. The baskets would be watched over from Chicago. Never mind that the majority of its revenues and real estate were and are in basket three. Or that Boeing's managers would now have the added challenge of flying all this blind, relying on remote readouts of the situation in Chicago instead of eyeballing it directly. The goal was to change Boeing's culture.

And in that, Condit and Stonecipher succeeded. In the next four years, Boeing's detail-oriented, conservative culture became embroiled in a series of scandals. Its rocket division was found to be in possession of 25,000 pages of stolen Lockheed Martin documents. Its CFO (ex-McDonnell) was caught violating government procurement laws and went to jail. With ethics now front and centre, Condit was forced out and replaced with Stonecipher, who promptly affirmed: 'When people say I changed the culture of Boeing, that was the intent, so that it's run like a business rather than a great engineering firm.' A General Electric alumnus, he built a virtual replica of GE's famed Crotonville leadership centre for Boeing managers to cycle through. When Stonecipher had his career-ending scandal (an affair with an employee), it was another GE alumnus - James McNerney Jr - who came in from the outside to replace him.

Walter James 'Jim' McNerney Jr. (*b* 22 August 1949) was a business executive who was chairman of The Boeing Company until 1 March 2016. He previously served as President and CEO of the company until July 2015. McNerney began his business career at Procter & Gamble in 1975, working in brand management. He worked as a management consultant at McKinsey from 1978 to 1982.

McNerney joined General Electric in 1982. There, he held top executive positions including president and CEO of GE Aircraft Engines and GE Lighting; president of GE Asia-Pacific; president and CEO of GE Electrical Distribution and Control; executive vice president of GE Capital, one of the world's largest financial service companies; and president of GE Information Services. McNerney competed with Bob Nardelli and Jeff Immelt to succeed the retiring Jack Welch as chairman and CEO of General Electric. When Immelt won the three-way race, McNerney and Nardelli left GE; McNerney was hired by 3M in 2001.

From 2001 to 2005, McNerney was chairman of the board and CEO of 3M, a

Right: Walter James 'Jim' McNerney Jr. who replaced Harry Stonecipher following the discovery of intimate e-mails in Stonecipher's e-mail account. It was McNerney who gave the MAX the go-ahead.

Below: Boeing's 'ivory tower' corporate HQ at 100 North Riverside, Chicago .

$20 billion global industrial company with leading positions in electronics, telecommunications, industrial, consumer and office products, health care, safety and other businesses.

On 30 June 2005, The Boeing Company hired McNerney as Chairman, President and CEO. McNerney oversaw the strategic direction of the Chicago-based, $61.5 billion aerospace company with a focus on spending controls. As Boeing's first CEO without a background in aviation he took the decision to upgrade the 737 series to 737 MAX instead of developing a new model.

As Richard Aboulafia Vice President of Analysis at the Teal Group and editor of their World Military and Civil Aircraft Briefing, a forecasting tool said at the time: 'You had this weird combination of a distant building with a few hundred people in it and a non-engineer with no technical skills whatsoever at the helm'. Even that might have worked - had the commercial-jet business stayed in the hands of an experienced engineer steeped in STEM disciplines. Instead, McNerney installed an MBA with a varied background in sales, marketing, and supply-chain management'. It shocked many long-term Boeing people.

Aboulafia again: 'The company that did not once speak finance was now, at the top, losing its ability to converse in engineering. It was not just technical knowledge that was lost, it was the ability to comfortably interact with an engineer who in turn feels comfortable telling you their reservations, versus calling a manager 1,500 miles away who you know has a reputation for wanting to take your pension away. It was a very different dynamic. As a recipe for disempowering engineers in particular, you could not come up with a better format.'

Jim Collins is a student and teacher of what makes great companies tick, and a Socratic advisor to leaders in the business and social sectors. Having invested more

The flight across America in the 737 that day took in some amazing sights - such as the Great Salt Lake...

...and the irrigated 'circles' of the farms in the mid-west. *(both author)*

than a quarter-century in rigorous research, he has authored or co-authored a number of books that have sold in total more than 10 million copies worldwide. As he wrote in 2000, 'If there's a reverse takeover, with the McDonnell ethos permeating Boeing, then Boeing is doomed to mediocrity. 'There's one thing that made Boeing great all the way along. They always understood that they were an engineering-driven company, not a financially driven company. If they're no longer honouring that as their central mission, then over time they'll just become another company.' It's now clear that long before the software lost track of its airliners' true bearings, Boeing lost track of its own.

Chapter Eleven

Taking it to the MAX.

A new version - or new airliner?

Boeing, launched their third-generation 737s to counter the A320. Airbus also developed the 320 into the 321, in competition with the 757.

The rivalry threatened to change even more in 2010 when Airbus introduced a version of the 320 called the NEO - for 'new engine option' - that offered large improvements in fuel efficiency, range and payload. The following year, American Airlines warned that it might abandon Boeing and buy hundreds of the new Airbus models. Boeing responded with a rush program to re-engineer the 737, modify the wings and make other changes to improve the performance of the airliner and give it some perceptible advantage over the A320NEO.

While officially, and publically downplaying the prospects for next-generation replacements for the A320 and 737, around 2006 Airbus and Boeing became increasingly engaged in studies, with formal initiatives launched on both sides.

According to industry insiders, Airbus called its A320 replacement study the NSR or New Short-Range aircraft. The NSR was baselined against the A320-200 for overall performance and cost terms and was aimed at a provisional service-entry date of 2012 or 2013. The NSR study, also referred to by its US competitor as the 'A-1', was broken into a three-phase effort. Phase 1 evaluated overall performance and operating cost targets, potential market assessments and technology candidates. Phase 2 was fine-tuning predicted improvements in fuel burn, emissions, noise and operating costs per hour and Phase 3 examined industrial implications concerned with partner workshares, technology development goals and airline advisory input.

Overall design concepts revolved around an all-composite primary fuselage and wing structure, more electric systems, advanced aerodynamics, including natural laminar flow and unstable, low-drag configurations, several cross-section options and integrated avionics with provision for enhanced vision systems (EVS).

Boeing called its 737 successor study the 737RS or Replacement Study, which was part of a project called 'Yellowstone'. Initial work on the Yellowstone 1 (Y1) was understood to represent just one of several possible replacement concepts and was one of the three Yellowstone new-generation studies emerging from the broad-based 'Project 20XX' advanced technologies initiative behind the Sonic Cruiser and subsequently the 787. Y1 was thought to cover the 100- to 200-passenger range; Y2, which became the 787, covered the 200- to 350-seat range; and Y3 covered the range for what could eventually become a successor to the 777-300/300ER.

According to sources, Boeing accelerated the pace of the 737RS study and planned to make its initial pass on prospective supplier teams by mid-2006. The RS/Y1 concept was based around an all-composite 787-like structure, fly-by-wire, more-electric system architecture, EVS-integrated avionics flight deck, and a cabin cross-section that was wider than A320. Aerodynamic improvements included a longer span wing, single-slotted flaps, raked and blended-winglet wingtip options, blended fin root and 787-like nose and flightdeck.

Initial results from both NSR and RS/Y1 studies were not encouraging. Acting completely independently, the two studies came up with similar results for their individual concepts, which fell far short of the targets set for the 2012 timeframe airliner.

Airbus NSR Phase 1 results were believed to have indicated that if all the then available and projected advanced technology was poured into the aircraft, the best specific fuel consumption reduction would be four per cent, the best operating cost reduction would be three per cent and the best emissions reduction would be five per cent. The numbers were said to be within 0.5-1% for all parameters for the initial phases of Boeing's RS/Y1.

Engines remained the key and the stumbling block to progress, as acknowledged by Boeing Commercial Airplanes vice-president sales Scott Carson, who said 'Right now, there is no engine. To build a 737 replacement without a next-generation engine would be a dreadful mistake for us to make'. Airbus chief executive Gustav Humbert also said that there were no imminent plans to create a new single-aisle aircraft to succeed the A320 family, with timing being dictated by market demand and the emergence of new engine technology. 'Both Boeing and Airbus will need an all-new engine, which the engine manufacturers say will not be ready until 2013-14, so the possible entry into service of such an aircraft could be possible around the middle of the next decade'.

Yellowstone was a Boeing Commercial Airplanes project to replace its entire civil aircraft portfolio with advanced technology aircraft. New technologies to be introduced included composite aerostructures, more electrical systems reducing reliance on heavier

BOEING 737 MAX-7

Length:	110 ft 4 in (33.63 metres)	**Fuel (max):**	6878 US gals (26035 lit)
Height:	40 ft 9 in (12.42 metres)	**Range:**	3825 miles (6155 km)
Span:	117 ft 9 in (35.90 metres)	**Cruise Speed:**	0.68 mach
Empty Weight:	-	**Pax Capacity:**	138-172
Max Weight:	177009 lbs (80,290kg)	**Powerplants:**	2 x CFM LEAP-1B

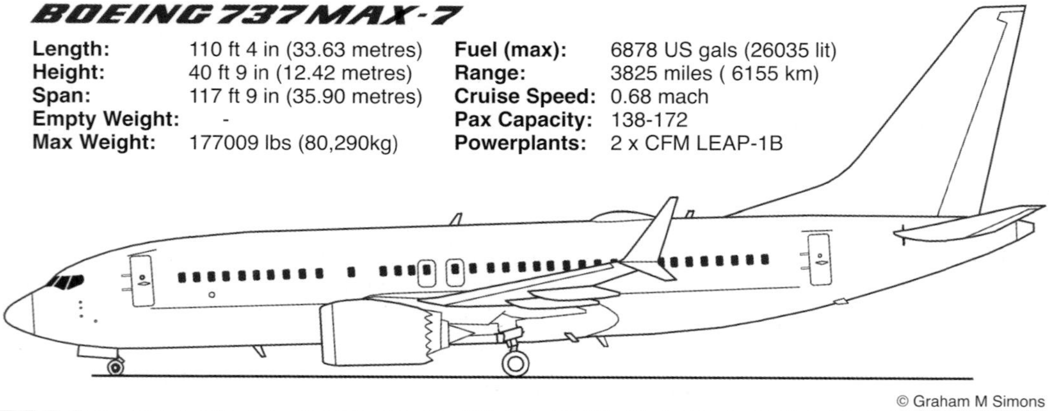

© Graham M Simons

Two views of the take-off of N87040 - the 737 MAX 8 demonstrator - at the 2016 Farnborough Air Show. The inset picture shows the very start of the roll, with engine fogging caused by sudden drop in air pressure as the engines spool up in both intakes of the LEAP 1B engines, and clouds of spray behind.

The main picture shows the airliner well into its take-off streaming clouds of spray behind - typical British airshow weather!
(both Kirk Smeeton Collection)

hydraulic systems, and more fuel-efficient turbofan engines such as the Pratt & Whitney PW1000G Geared Turbofan, General Electric GEnx, the CFM International CFM56, and the Rolls-Royce Trent 1000.

The term 'Yellowstone' referred to the technologies, while 'Y1' through 'Y3' designators referred to the actual aircraft. The first of these projects, Y2, entered service as the Boeing 787.

The Boeing Y1 was to replace the Boeing 737, 757, and 767-200 product lines. The Y1 covered the 100- to 250-passenger market, and was to be the second Yellowstone Project airliner developed. Boeing submitted a patent application in November 2009, that was released to the public in August 2010, which envisioned an elliptical composite fuselage, and appeared to signal the company's planning for the 737 successor. In early 2011, Boeing outlined plans for a 737 replacement that would arrive in 2020.

In July 2011, eight months after Airbus announced its LEAP-X re-engined A320 NEO, Boeing announced the 737 MAX, an updated and re-engined version of the 737 NG, rather than progress further with Y1 concept. It looked like a hasty move to secure a large

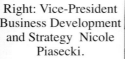

Left: Boeing Commercial Airplanes CEO Scott Carson.

Right: Vice-President Business Development and Strategy Nicole Piasecki.

order from American Airlines, which was eventually split between the two manufacturers. The new family was named 737 MAX-7, -8 and -9 to harmonise the brand. The first aircraft made its first flight on 29 January 2016.

To keep costs down, as with all previous iterations, the redesign had to lie within the original 1968 FAA certification of the type and not be looked upon officially as a new airliner. Airbus had similar requirements for the NEO. In its marketing literature comparing the MAX to the earlier Next-Generation 737, Boeing wrote: 'same pilot type rating, same ground handling, same maintenance program, same flight simulators, same reliability.'

Part of the delay was waiting for new technologies in engines, aerodynamics, materials and other systems to be developed. But part was also due to the 737NGs strong order book. Boeing Commercial Airplanes CEO Scott Carson insisted in 2008 that 'the effort to develop a 737 replacement has not been abandoned, only pushed out to ensure that what results has a long market life'.

Once again, in order to save vast amounts of time and money, Boeing decided to certify the MAX series under an amended type certificate as part of the 737 family rather than go for a whole new type certificate for a new airliner.

During the press conference announcing the re-engined 737, Nicole Piasecki, vice president of business development and strategy, explained why Boeing chose the MAX name: 'We wanted the name to capture how exceptional the 737 was not only in terms of its performance, but we wanted it to be able to differentiate the 7, 8 and 9. We wanted to make sure the name was easily identifiable from four-year olds up to ninety-year olds, and we wanted to make sure that it represented the best that it will truly be. We thought about how do you convey superiority, the best, the gold standard in single-aisle airplanes. And how do you come up with a name to describe already a great airplane. We wanted to make sure that it talked about what it was going bring to the industry in terms of maximum benefit, maximum competitive advantage for our customers, maximum value and absolute maximum in what an airplane could deliver to our customers. So we came up with something that fit that, and we will be calling this airplane the 737 MAX'.

The LEAP® (Leading Edge Aviation Propulsion) powerplant was a high-bypass turbofan produced by CFM International, a 50-50 joint venture between American GE Aviation and French Safran Aircraft Engines, formerly called Snecma. The engine was a successor of the successful CFM56 competing with the Pratt & Whitney PW1000G to power narrow-body aircraft. The engine was officially launched as LEAP-X on 13 July 2008.

The LEAP-1B® had been selected by Boeing as the exclusive powerplant for the new 737 MAX. It offered MAX operators exceptional technical, economic and environmental performance, with a fifteen per cent reduction in fuel consumption and carbon dioxide emissions versus current engines, and up to fifty per cent margin on nitrogen oxide emissions versus CAEP/6 Standard, and in compliance with the most stringent noise standards, the ICAO's Chapter 14 regulations

It was a new 69.4 inch diameter engine with eighteen woven carbon-fibre fan blades giving a bypass ratio of 9:1 versus 5.1:1 for the CFM56-7. The rated thrust of the LEAP-1B28 was 29,317 pounds. The turbine had flexible blades manufactured by a resin transfer moulding process, which were designed to untwist as the rotational speed increased. This, along with advanced hot-section materials, delivered an overall pressure ratio of 41:1, compared to 28:1 for the CFM56-7. The engine was fifteen per cent more fuel-efficient than the CFM56-7B.

The sawtooth pattern or 'chevrons' on the trailing edges of the fan nozzles were developed by NASA to smooth the mixing of the fan and core air flows. This reduced turbulence giving a significant reduction in noise.

On 12 February 2012 Boeing announced that the final phase of wind tunnel testing, a major milestone in 737 MAX programme airliner family development, would begin within the next seven days. 'Wind tunnel testing is on the critical design path of the program,' said Michael Teal, chief project engineer and deputy program manager, 737 MAX program. 'Based on previous work in the wind tunnel, we are confident this final phase of testing will substantiate our predictions of the aerodynamic performance of the airplane'.

BOEING 737 MAX-8
Length: 129 ft 6 in (39.5 metres)
Height: 40 ft 4 in (12.3 metres)
Span: 117 ft 10 in (35.9 metres)
Empty Weight: -
Max Weight: 181,200 lbs (82,191 kg)
Fuel (max): 6,820 US gals (25,817 lit)
Range: 3515 miles (6510 km)
Cruise Speed: 0.79 mach
Pax Capacity: 162 - 210
Powerplants: 2 x CFM LEAP-1B
© Graham M Simons

'Testing will begin at QinetiQ's - pronounced 'kin-et-ik' - test facility in Farnborough, in the UK, where engineers intend to substantiate the forecasted low-

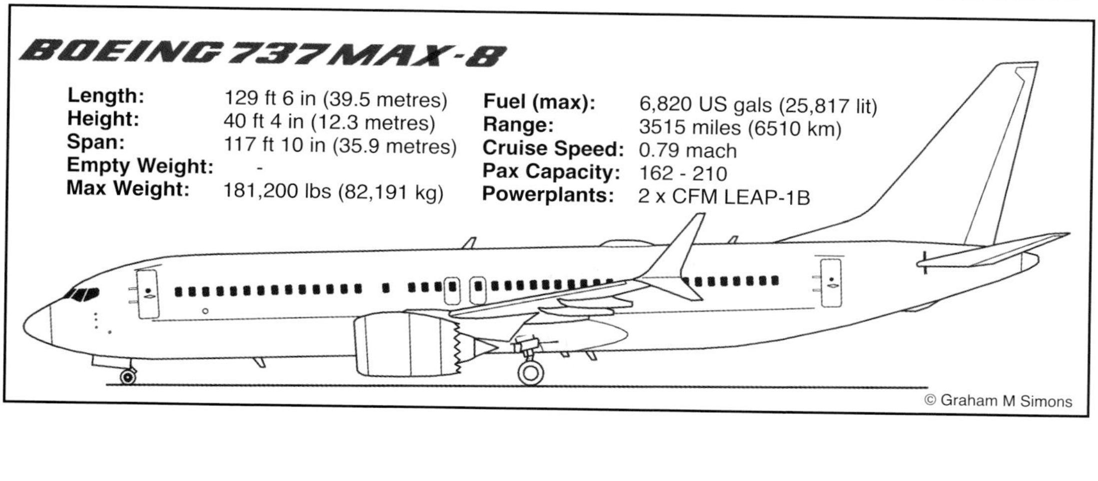

N87040 - the MAX-8 demonstrator - returns to earth after a spirited, if totally unrealistic, display at Farnborough that included near vertical climbs, negative G and steep turns that an airliner would never perform in service. *(Kirk Smeeton Collection)*

speed performance of the 737 MAX on take-off and landing. QinetiQ uses its domain knowledge to provide technical advice to customers in the global aerospace, defence and security markets'.

'Testing also will be completed at the Boeing Transonic Wind Tunnel in Seattle to substantiate the forecast of the high-speed performance of the airliner'.

The models used for Next-Generation 737 wind tunnel testing, with modifications made to the aft fuselage, struts and nacelles, in addition to the new engine, were used for the tests. Test completion in mid-2012 was a major step toward the final configuration of the 737 MAX.

'This final phase of wind tunnel testing confirms that we are on track to complete our design goals and deliver the 737 MAX to customers beginning in 2017,' said Teal. The 737 MAX was now announced as a new engine variant of the world's best-selling airliner that built on the strengths of the Next-Generation 737.

The 737 MAX incorporates the latest-technology CFM International LEAP-1B engines to deliver the highest efficiency, reliability and passenger appeal.

Boeing was claiming that airlines operating the 737 MAX would see a ten-to-twelve per cent fuel burn improvement over today's most fuel-efficient single-aisle airliners and a seven per cent operating cost per-seat advantage over future competition.

Just over a month later, on 26 March,

BOEING 737 MAX-9

Length:	138 ft 4 in (42.16 metres)	**Fuel (max):**	6,820 US gals (25,817 lit)
Height:	40 ft 4 in (12.3 metres)	**Range:**	3515 miles (6510 km)
Span:	117 ft 10 in (35.9 metres)	**Cruise Speed:**	0.79 mach
Empty Weight:	-	**Pax Capacity:**	162 - 210
Max Weight:	181,200 lbs (82,191 kg)	**Powerplants:**	2 x CFM LEAP-1B

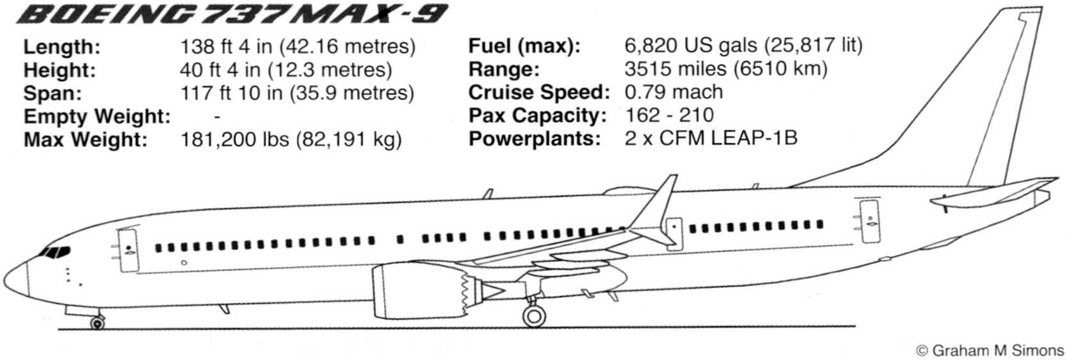

N7379E, the Max-9 demonstrator, comes in to land after its display at the 2017 Paris Air Show. The number '167' on the nose is part of the traditional participants numbering for the show. The airliner was later re-worked and delivered to Lion Air. *(Kirk Smeeton Collection)*

it was reported that Boeing Business Jets had already begun talking to potential customers about a VIP variant of the re-engined 737 MAX - called the BBJ MAX - for 2018-19.

Captain Steve Taylor, president of Boeing Business Jets, said: 'We are working with several existing customers and one would like to be the launch customer for that aircraft.'

The Leap-1B-powered MAX began to roll off the production line in 2017. Taylor noted that a BBJ customer typically wanted a much shorter lead-time for delivery post-completion than Boeing could currently offer for the BBJ MAX 'so it takes a fairly special customer to be ready to commit to 2018 or 2019'.

He said: 'There are several long-term BBJ customers who would be ideal buyers for it ,and we have started those conversations.'

Taylor went on record saying that the BBJ 2 MAX - the VIP equivalent of the 737-8 - would offer a potential range of 6,200 nauticial miles.

Progress was swift. On 11 April 2013 it was revealed that Boeing had made a series of design updates to the 737MAX to optimise the new-engine variant's performance further.

'The 737 MAX is on-track to deliver substantial fuel-savings to customers starting in 2017,' said Beverly Wyse, vice president and general manager, 737 programme. 'We've made several design decisions that support the performance targets for the MAX and evolve the Next-Generation 737's design within the scope of the 737 MAX program.'

'Those design decisions included:

- 'Aft body aerodynamic improvements: The tail cone will be extended and the section above the elevator thickened to improve steadiness of airflow. This eliminates the need for vortex generators on the tail. These improvements will result in less drag, giving the airplane better performance'.
- 'Engine installation: The new CFM International LEAP-1B engines will be integrated with the wing similar to the aerodynamic lines of the 787 Dreamliner engine with its wing. A new pylon and strut, along with an eight-inch nose gear extension, will maintain similar ground clearance to today's 737 while accommodating the larger engine fan. The nose gear door design is altered to fit with this revision.'
- 'Flight control and system updates: The flight controls will include fly-

American Airlines N324RA - a 737 MAX-8. *(Matt Black collection)*

by-wire spoilers, which will save weight by replacing a mechanical system. The MAX also will feature an electronic bleed air system, allowing for increased optimisation of the cabin pressurisation and ice protection systems, resulting in better fuel burn'.

Other changes to the airliner included strengthening the main landing gear, wing and fuselage to accommodate the increase in loads due to the larger engines. Boeing continued to conduct aerodynamic, engine and airliner trade studies as the team works to optimise the design of the aircraft by mid-2013.

'We also continue to do work in the wind tunnel to affirm the low- and high-speed performance of the 737 MAX design,' said Michael Teal, chief project engineer and deputy program manager, 737 MAX programme. 'Based on design work and preliminary testing results, we have even more confidence in our ability to give our customers the fuel savings they need while minimising the development risk on this program.'

'A possible revision to the wingtips on the MAX also is being tested in the wind tunnel to see if this new technology could further benefit the airplane.

'Any new technology incorporated into the MAX design must offer a substantial benefit to our customers with minimal risk for the team to pursue it', said Teal. 'On the 737 MAX we are following our disciplined development

Left: Michael Teal, chief project engineer and deputy program manager, 737 MAX programme.

Beverly Wyse, general manager 737 programme.

process and continue to work on an airplane configuration that will provide the most value for our customers.'

Clearly, Boeing's goal with the newest version of the 737 was to keep any change to a minimum; but as the company closed in on the final design, it appeared that more substantial upgrades were in the works to meet performance targets.

The company said it would adopt an eight-inch nose-gear extension to provide adequate ground clearance for the CFM Leap-1B engine. The decision backed up earlier disclosures that Boeing and CFM were considering increasing the fan diameter to 69 inches when the 737 MAX's firm configuration was completed around mid-2013.

The MAX would be fitted with fly-by-wire spoilers; an option first studied for the 737 Next Generation Plus - a proposed interim machine between the then-current model and a clean-sheet New Small Airplane - later abandoned. Boeing explained that the fly-by-wire spoiler system, which dated to the 1980s on the 757/767, would be mainly used for weight savings. However, as the newer system was less mechanically complex, it was expected to reduce maintenance costs and provide potential load-alleviation benefits.

The aft-fuselage changes originally outlined by Boeing in August 2011 may also have been more substantial than first suggested. The revision of the Section 47/48 tail cone to a lower-drag shape similar to the 787's was also expected to include the elimination of an aft-body join and changes to the aft pressure-relief port cavity so as to reduce drag. At the time, Boeing would not confirm these details, stating that more changes would be visible when updated impressions of the MAX were released around the Farnborough air show in July. Beverly Wyse further explained: 'We've made several design decisions that support the performance targets for the MAX and evolve the 737 NG's design within the scope of the 737 MAX program.'

Despite suggestions made by Airbus that Boeing's design fell short of the targets set for fuel burn, Wyse went on to insist that '...the 737 MAX is on track to deliver substantial fuel savings to customers starting in 2017.'

Which version do you want?

Boeing offered potential customers the MAX-7, a slightly shorter, longer-range version of the MAX-8 but longer than the 737-700. Overall length was 110 feet 4 inches, with a maximum capacity of 172 passengers, hence the two pairs of

N7201S, the first MAX-7, intended for Southwest Airlines, seen on take off at the 2018 Farnborough Air Show. *(Kirk Smeeton Collection)*

Icelandair's 737 MAX-8 TF-ICU is caught by the camera about to land at Brussels Airport.
(Emma Dayle collection)

overwing exits. Its range was approximately 3825 nautical miles, and it was initially expected to be in service sometime in 2019.

However, in April 2016, two years before the first one was built, it was announced that the 737 MAX-7 would be stretched by two seat rows, due to sales of the 737 MAX-7 being sluggish, winning just sixty orders. This was achieved by inserting a forty-six inch fuselage plug forward of the wing and a thirty inch fuselage plug aft of it. Sales were so dire that Boeing considered cancelling the series.

The underlying problem with short versions of airliners is that they have proportionally more weight per seat, meaning a greater fuel burn with less potential for yield. Looking back at all generations of the 737, the shortest models have been the least successful:

737-100 - 30 sold out of 1144 originals; 2.6 per cent of the build.

737-500 - 389 sold out of 1988 classics; 19.6 per cent of the build.

737-600 - 69 sold out of approximately 7100 NGs; 1.0 per cent of the build.

737-MAX7 - 63 sold out of approximately 3600 MAXs; 1.7 per cent of the build.

So Boeing had positioned the -7 as the long-range MAX, less fuselage length and passengers meant less weight, and with the -8 strength through the use of thicker guage aluminium wing skins, it could still lift a lot of fuel.

The 737 MAX-7 seating capacity was 153 in two-class configuration or 172 in high density. The range was 3825 nautical miles by using stronger MAX-8 wings and landing gear, making it the longest range MAX.

Boeing claimed the MAX-7 would carry twelve more passengers 400 nautical miles further and with seven per cent lower operating costs per seat than the A319NEO.

Then, on 18 December 2017, Boeing and Bombardier of Canada clashed at US International Trade Commission over pricing. Kevin McAllister, the head of the Boeing Commercial Aircraft division, argued that Bombardier's sale of its C Series jets at what he said were below fair-market prices posed an existential threat to the 737 MAX-7. 'Our Max-7 is at extreme risk. If you don't level the playing field now, it will be too late'.

Boeing wanted to have tariffs imposed on sales of the C Series to compensate for what it described as

SpiceJet Limited is a low-cost airline headquartered in Gurgaon, India. It is the second largest airline in the country by number of domestic passengers carried. The airline operates 630 daily flights to 64 destinations, comprising 54 Indian and 10 international destinations from its hubs at Delhi, Kolkata, Mumbai and Hyderabad. Here VT-MXC, a 737 MAX-8 is seen coming in to land. *(Matt Black Collection)*

below fair-market pricing. This would directly impact deliveries of an order of seventy-five C Series jets ordered by Delta in 2016. Delta contended that Boeing did not lose this sale to Bombardier: 'When we chose to add the CS100 aircraft to our fleet, Boeing did not have the right-sized aircraft.'

The first flight of the MAX-7 was on 16 March 2018 with N7201S. It was flown by Boeing Test and Evaluation Captains Jim Webb and Keith Otsuka from Renton to Boeing Field and lasted three hours and five minutes.

Keith Leverkuhn, vice president and general manager 737 MAX programme, commented: 'Everything we saw during today's flight shows that the MAX-7 is performing exactly as designed. I know our airline customers are going to enjoy the capabilities this aeroplane will bring to their fleets'.

The 737 MAX -8

This was to be the baseline model, equivalent of the 737-800 and was the first variant developed in the 737 MAX series. It replaced the 737-800 with a longer fuselage than the MAX 7. Boeing planned to improve its range from 3,515 nautical miles to 3,610 nautical miles after 2021. On 23 July 2013, Boeing

BOEING 737 MAX-10

Length:	143 ft 8 in (43.8 metres)	**Fuel (max):**	TBA
Height:	40 ft 4 in (12.3 metres)	**Range:**	TBA
Span:	117 ft 10 in (35.9 metres)	**Cruise Speed:**	0.79 mach
Empty Weight:	TBA	**Pax Capacity:**	188-204
Max Weight:	TBA	**Powerplants:**	2 x CFM LEAP-1B

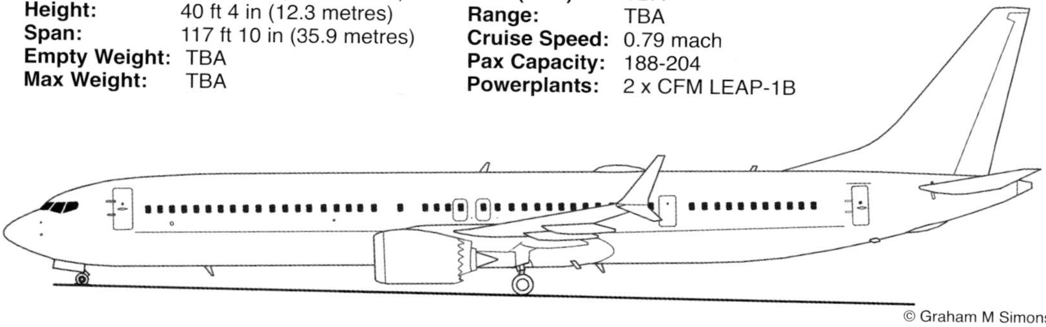

© Graham M Simons

The first MAX -10 rolled out to a 'Boeing Family' event on 22 November 2019. *(BJ)*

completed the firm configuration for the 737 MAX-8, which had a lower empty weight and higher maximum takeoff weight than the A320NEO.

The Boeing 737 MAX 8 completed its first flight testing in La Paz, Bolivia. The 13,300-foot altitude at El Alto International Airport tested the MAX's capability to take off and land at high altitudes. Its first commercial flight was operated by Malindo Air on 22 May 2017, between Kuala Lumpur and Singapore as Flight OD803.

The 737 MAX 9

This was to replace the 737-900 and had a longer fuselage than the MAX 8. Boeing planned to improve its range from 3,510 nautical miles to 3,605 nautical miles after 2021. Lion Air was the launch customer with an order for 201 in February 2012. It made its roll-out on 7 March 2017, and first flight on 13 April 2017; it took off from Renton Municipal Airport and landed at Boeing Field after a two hour forty-two-minute flight.

Boeing 737-9 flight tests were scheduled to run through 2017, with thirty per cent of the -8 tests repeated; aircraft 1D001 was used for auto-land, avionics, flutter, and mostly stability-and-control trials, while 1D002 was used for environment control system testing. It was certified by February 2018. Asian low-cost carrier Lion Air Group took delivery of the first on 21 March 2018, before entering service with Thai Lion Air.

Kevin McAllister, president and CEO of the commercial airplanes unit.

The 737 MAX -10

This was a stretched MAX-9 due to slow sales against the A321NEO and was expected to be in-service in 2020. Boeing launched the longest version of the MAX on 19 June 2017, to compete with the Airbus 321NEO and fill the gap in the market left by the 757. At the time of writing, the MAX-10 has over 361 firm orders and commitments.

Boeing's Randy Tinseth said that the 737 MAX-10 would offer the same capacity as the slightly longer Airbus A321NEO, but would be 2.8 tons lighter, so having lower fuel consumption and slightly greater range. He went on to say that 'We aren't looking to simply build something on par with the A321NEO. We're bringing a better airplane to the market - and that's our focus.'

The first MAX -10

fuselage arrived at Renton in early April 2019.

Boeing rolled out its latest and largest 737 Max variant, the 737-10, at its Renton plant on 22 November 2019. Rollouts often make for great marketing events with media, airline customers, and even elected officials on hand. But this unveiling was much more low-key, just a Boeing family affair. There was no media circus, no fanfare, just hundreds of Boeing employees turning out for the official rollout of the company's biggest 737 MAX - a traditional rite in the birth of an airliner that was more muted this time, due to the then eight-month-long grounding of all 737 MAX airliners. The aircraft, which was more than a simple stretch as it featured a revised main landing gear design, was expected to fly in 2020.

Longer legs?

The MAX-10 seats up to 230 passengers and is around five feet three inches longer than the MAX-9. The longer fuselage potentially required taller landing gear to ensure the appropriate clearance between the rear fuselage and the ground during take-off rotation.

Although the engine was unchanged, the fuselage was lengthened so the main landing gear had to be modified to enable adequate clearance of the longer body for rotation on takeoff and landing and to ensure the aircraft remained stall rather than pitch-limited.

Boeing's original plan was to lengthen the main landing gear. However, since the fan diameter was unchanged an easier solution which required no changes to the wheel well was adopted, namely a 'semi-levered' design which was more

The three 'positions' of the MAX-10 trailing link landing gear, with the direction of travel towards the right - showing how the leg 'grows' before take-off and then reduces in length to fit into the landing gear wells.

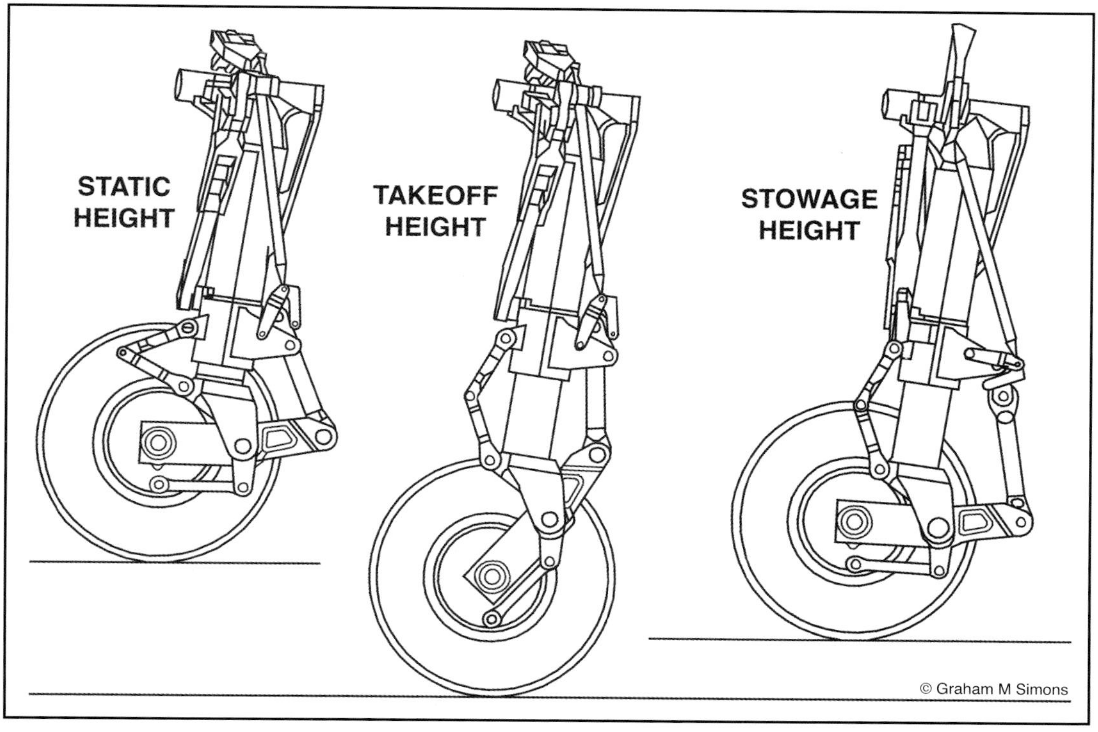

Norwegian's 737 MAX-8 EI-FYB is about to get airborne from Edinburgh. On the tail is Thomas Crean, an Irish seaman and Antarctic explorer who was awarded the Albert Medal for Lifesaving. Crean was a member of three expeditions to Antarctica, including Robert Falcon Scott's 1911-13 Terra Nova Expedition. *(Jim Patterson Collection)*

commonly known as a trailing-link, similar to that used on the 777-300ER and 787-10, that shifted the rotation point slightly aft. The gear was also telescopic and contracted during retraction to fit into the existing wheel well so preserving commonality within the MAX family by allowing the landing gear to be accommodated into the same wheel well as other MAX variants.

Details of the MAX -10 landing gear were released in a video by Boeing in late August 2018 in which the 737 MAX chief product engineer, Gary Hamatani, stated that Boeing had '...settled on a levered design that enables the gear to extend by nine and a half inches upon rotation during takeoff'. In addition to the lever system, the Max 10's main gear had what the company called an 'innovating shrinking mechanism' - also termed a 'shrink-link' made of steel that compressed the inner cylinder as the gear retracted, enabling it to fit in the same wheel well. From a pilot's perspective, there was absolutely nothing different from the MAX 10 landing gear and the existing Max family.

The problems created by the landing gear length are explained in detail in a later chapter.

Chapter Twelve

Two MAX Crashes

Lion Air Flight 610

This was a scheduled domestic flight operated by the Indonesian airline Lion Air from Soekarno–Hatta International Airport in Jakarta to Depati Amir Airport in Pangkal Pinang. On 29 October 2018, the Boeing 737MAX 8 operating the route crashed into the Java Sea 13 minutes after takeoff, killing everyone on board. The aircraft, registration PK-LQP, line number 7058, was powered by two CFM International LEAP engines. It was leased from China Minsheng Investment Group (CMIG) Aviation Capital and had been delivered new to Lion Air on 13 August 2018. At the time of the accident, the aircraft had flown about 800 hours in service.

It was the first major accident involving the new Boeing 737 MAX series aircraft, introduced into service on 22 May 2017, and the deadliest involving the Boeing 737 Series, surpassing Air India Express Flight 812 in 2010. It was also the deadliest accident in Lion Air's eighteen-year history, surpassing the 2004 crash in Surakarta that killed twenty-five, and the second deadliest aircraft accident in Indonesia behind Garuda Indonesia Flight 152.

There were 189 people on board the aircraft: 181 passengers, consisting of 178 adults, one child and two infants, as well as six cabin crew and two pilots. The flight's cockpit crew were Captain Bhavye Suneja, a 31 year old Indian national who had flown with the airline for more than seven years and had about 6,028 hours of flight experience - including 5,176 hours on the Boeing 737 - and Indonesian First Officer Harvino, who like many Indonesian nationals, had only a single name. He had 5,174 hours of flight experience, 4,286 of them on the Boeing 737. The six flight attendants were also Indonesian.

The aircraft took off from Jakarta at 6:20 a.m. local time and was scheduled to arrive at Depati Amir Airport in Pangkal Pinang at 7:20 a.m. It took off in a westward direction before turning to establish a northeast heading, which it held until crashing offshore northeast of Jakarta in waters estimated to be up to one hundred and fifteen feet deep. The flight crew had requested clearance to return to the Jakarta airport nineteen nautical miles into the flight. The accident site was located eighteen nautical miles off the coast of the island of Java.

Communication between ATC and Flight 610 was suddenly lost at 6:33 am. ATC immediately informed authorities about the incident and the Indonesian National Search and Rescue Agency deployed three ships and a helicopter to the area. At 7:30 am, the agency received reports that Flight 610 had crashed close to an offshore oil platform. Workers on the platform reportedly saw the aircraft crash with a steep nose-down angle. Boats from the platform were immediately deployed, and debris from the crashed aircraft was found shortly after.

On 29 October, Indonesia's Transportation Ministry ordered all of the country's airlines to conduct emergency inspections on their 737 MAX 8 aircraft. The ministry also launched an audit on Lion Air to see if there were any problems with its management system. The Transportation Ministry announced that all Indonesian Boeing 737 MAX-8 aircraft were airworthy and were allowed to resume normal operations on 31 October.

A spokesperson for the Indonesian Basarnas, the National Search and Rescue Agency, confirmed to reporters that the aircraft had crashed. Muhammad Syaugi, head of Basarnas, later confirmed that there had been casualties, but did not say how many.

In response to the crash, the Indonesian Transportation Ministry set up crisis centres in Jakarta and Pangkal Pinang. Lion Air also offered free flights for the families of the victims to Jakarta. On 30 October, more than ninety relatives were flown to Jakarta to identify the victims. CEO of Lion Air, Edward Sirait stated that accommodation had been provided for the relatives, and later added that relatives should go to Halim Perdanakusuma International Airport for further information. The Indonesian National Police announced that a trauma healing team would be provided for the relatives of the victims.

Indonesia's People's Representative Council announced on 29 October that they would examine the Standard Operating Procedure of Lion Air and the airworthiness of the aircraft. They would also examine the health history of the crew of Flight 610. The Speaker of the People's Representative Council, Bambang Soesatyo, later asked the government to enforce stricter rules for the aviation industry and to audit every airliner in the country. On 1 November, Indonesian Minister of Transportation Budi Karya Sumadi announced that the government would evaluate every low-cost carrier in Indonesia.

As twenty of the passengers were employees of the Indonesian Ministry of Finance, Sri Mulyani, the Indonesian Finance Minister, immediately visited the Indonesian Search and Rescue Agency's office in Jakarta, seeking coordination and further information. She later announced that every employee of her ministry should wear a black ribbon for a week to commemorate the victims. Posthumous awards would be given to the twenty victims and scholarships would be given to the victims' children.

Minister of Health Nila F Moeloek and Minister of Transportation Budi Karya Sumadi visited the relatives of the victims. Indonesian President Joko Widodo, who was attending a conference in Bali at the time of the crash, visited the recovery efforts at the Port of Tanjung Priok the next day.

One immediate result of the crash was that the Australian Department of Foreign Affairs and Trade announced that its staff would be banned from flying on Lion Air, as well as its subsidiary airlines Batik Air and Wings Air, until the cause of the accident was known. The Indonesian Minister of Transportation, Budi Karya Sumadi, later stated that his ministry would hold talks with the Australian government about the warning.

The government-owned social insurance company Jasa Raharja announced that the victims' families would each receive fifty million rupiah, or just over US$3,700, in compensation.

A side effect of the crash was the severe criticism of the media for their coverage of the event. In the aftermath of the crash, the Indonesian media were warned by the Indonesian Broadcasting Commission about their unethical coverage of the crash. Some media were accused of pushing the family members to answer unethical questions. Chairman of Indonesia's Alliance of Independent Journalists, Abdul Manan, stated that images of debris were broadcast repeatedly and inappropriately. This, in turn, traumatised relatives of the victims. In response, some Indonesians voiced their anger and disappointment on social media. One online media outlet, OkeZone, posted a video showing a large mob of reporters surrounding a visibly shaken woman outside the Lion Air Crisis Centre in Jakarta and rapidly firing questions at her from all directions. 'What would happen if they don't find the plane?' one reporter asked. 'How did you feel after hearing about the crash?' asked another. In a separate video from another local media company, reporters could be heard asking relatives of another victim: 'Did you foresee something like this happening?' 'Did you have a premonition this would happen?' another shouted. Ahmad Arif - a journalist at Kompas and author of the book Disastrous Journalism; Journalism Disaster - said that aggressive eliciting of emotion from vulnerable interviewees was a common practice among journalists in Indonesia. 'Those kind of questions are asked by news hunters to stir their feelings, often deliberately exploiting them just for the sake of dramatising already distressful unfortunate events'.

Left: CEO of Lion Air, Edward Sirait addresses the world's media following the crash of Lion Air Flight 610.

Below: PK-LQP, the Boeing 737 MAX-8, seen before the crash.

On 31 October, Transportation Minister Budi Karya Sumadi temporarily suspended Lion Air's technical director Muhammad Arif from his duties based on the crash investigation. Budi said the ministry had also suspended an unspecified number of Lion Air technicians who had cleared the aircraft for its final flight.

Recovery

Indonesia's National Search and Rescue Agency launched a search and rescue operation, with assistance from the Indonesian Air Force, the Indonesian Navy, and the Republic of Singapore Navy. Basarnas dispatched about 150 people in boats and helicopters to the site of the accident. Civilian vessels also responded to the reports of a downed aircraft, and the crew of a tugboat reported to authorities in Tanjung Priok that they had witnessed an aircraft crash at 6:45 a.m. and located debris in the water at 7:15 a.m. The Indonesian Agency for Assessment and Application for Technology deployed the research ship Baruna Jaya, which had been previously used in the search for Adam Air Flight 574 and Indonesia AirAsia Flight 8501.

Officials from the Indonesian National Search and Rescue Agency announced that the search and rescue operation would be conducted for seven days and be extended by three days if needed. A command centre was set up in Tanjung Priok. On 29 October, authorities said that all on board were presumed dead and that the first human remains had been recovered. Divers had located fragments of the aircraft's fuselage and assorted debris, but had yet to find the onboard flight recorders. An official from the Indonesian National Armed Forces suggested that most of the victims were still inside the fuselage, as in the days following the crash rescue personnel only managed to recover a small number of body parts. Officials stated that bad visibility and strong sea current hampered the search and rescue effort.

On the same day, the Indonesian National Search and Rescue Agency published the area of the search and rescue operation. It was divided into two main areas. On 30 October, the search area was further divided into thirteen sectors and widened, reaching as far as Indramayu to the east. Approximately 850 personnel from the Indonesian National Search and Rescue Agency, Indonesian National Armed Forces and volunteers participated in the operation. At least thirteen bodies were retrieved from the crash site. Indonesian officials confirmed that faint pings had been heard in the search area. On 31 October, it was reported that acoustic 'pings' had been detected, less than two miles from the group of eight current search points, which were possibly from one or both of the underwater locator beacons (ULBs) attached to the aircraft's flight recorders.

The first victim was identified on 31 October. At the time, more than a dozen body parts had been found by authorities. Some of the parts had drifted nearly three miles in the sea current. Police also reported that 152 DNA samples had been collected from the victims' relatives.

Hundreds of pieces of the aircraft had also been recovered; all of them were transported to Tanjung Priok, Jakarta. Authorities stated that the search area for dead bodies and debris would be focused in the sea off Karawang Regency, a coastal area of Java close to the crash site, as analysis showed that the sea currents in the area would bring debris to the south. A command centre was set up in Tanjung Pakis, Karawang to oversee the salvage effort. On the same day, authorities widened the search area from ten to fifteen nautical miles. In all, thirty-nine ships - including four equipped with sonar - and fifty divers were deployed to the search area. The Indonesian National Police announced that 651 personnel had joined and assisted in the search and rescue operation. Officials stated that the operation, starting from 31 October, would focus on finding the fuselage of the aircraft and the flight recorders.

The joint search and rescue team announced on 31 October that at least three objects, one of which was suspected to be one of the aircraft's wings, were found in the search area. On 1 November, searchers announced that they had found Flight 610's flight data recorder (FDR), which was located at a depth of one hundred and five feet. The cockpit voice recorder (CVR), however, was reported as not yet found. According to a transport safety official, authorities did not initially know for certain whether the 'crash survivable memory unit' was from the flight data recorder or cockpit voice recorder, as portions of it were missing. Haryo Satmiko, deputy chief of Indonesia's National Transportation Safety Committee (Indonesian: Komite Nasional Keselamatan Transportasi), the body investigating the crash, told journalists that the device's poor condition was evidence of the 'extraordinary impact' of the crash, which had separated the memory unit from its housing - an event that was not supposed to be able to happen. Despite the damage, investigators were able to recover data from the aircraft's most recent nineteen flights spanning sixty-nine hours and planned to begin analysis on 5 November.

A Boeing technical and engineering team, and a team from the US National Transportation Safety Board arrived on 31 October to help with the investigation being conducted by the NTSC. Personnel from the US Federal Aviation Administration and engine manufacturer GE Aviation were also sent to Indonesia. A team from Singapore, that had already arrived on the night of 29 October, was to provide assistance in recovering the aircraft's flight recorders. The Australian Transport Safety Bureau sent two of its personnel to assist with the downloading process of the FDR.

On 2 November, the joint search and rescue team deployed more than 850 personnel and forty-five vessels to the crash site. The aerial search area was widened to one hundred and ninety nautical miles and the underwater search area was widened to two hundred and seventy nautical miles. The joint search and rescue team announced that some engine parts were found in the search area. One of the aircraft's landing gears was recovered in the afternoon. Meanwhile, the Disaster Victim Identification team stated that at least 250 body parts had been recovered from the crash site.

On 3 November, it was reported that a volunteer Indonesian rescue diver had died during the search, on the afternoon of 2 November. It was believed he died from decompression sickness.

A second landing gear and both of the aircraft's engines were recovered by search and rescue personnel, and the main body of the aircraft had been located. The main wreckage of the aircraft was located seven and a half miles from the coast of Tanjung Pakis and was about two hundred and twenty yards from the location where the FDR was discovered. Divers were immediately dispatched to the area, where faint 'pings' from the ULB attached to the aircraft's CVR were also heard.

On 4 November, the Head of the Indonesian National Search and Rescue Agency Muhammad Syaugi announced that the search and rescue operation would be extended for another three days. Then, on 10 November the NSRA ended its search for victims, but on 22 November were continuing to search intensively for the CVR.

On 23 November, investigators concluded the victim identification process. Out of 189 people on board, 125 (89 men and 36 women) were identified, including the two foreigners. Sixty-four bodies were unaccounted for.

On 14 December, it was reported that Lion Air had paid US$2.8 million for a second attempt to search for the CVR, with a specialised boat being brought in to assist in the search, expected to last 10 days. The attempt began on 20 December, but ended on 3 January 2019 without success.

The NTSC announced plans to launch their own search, and were negotiating with the Indonesian Navy to borrow one of their ships. On 8 January 2019, they announced they would be resuming the search for the CVR, which was finally found and recovered by the Indonesian Navy on 14 January 2019, more than two months after the accident. Its location was near the crash site, at a depth of ninety-eight feet but covered by mud twenty-six feet thick.

Previous flight problems, Speculation and Flight Abnormalities.

The aircraft was used on a flight from Ngurah Rai International Airport, Bali to Soekarno-Hatta International Airport, Jakarta the night before the crash. Detailed reports from that flight revealed that the aircraft had suffered a serious incident, which left many passengers traumatized. Passengers in the cabin reported heavy shaking and a smell of burnt rubber inside the cabin. At one point, the aircraft had dropped more than two hundred feet in a few seconds. The seat belt sign was never turned off from takeoff to landing. A recording of air traffic control communications indicated that the pilot had called a 'mayday', but the crew later decided to cancel the call and continued the flight to Jakarta.

The aircraft's maintenance logbook revealed that the aircraft suffered an unspecified navigation failure on the captain's side, while the First Officer's side was reported to be in good condition. Passengers recounted that the aircraft had suffered an engine problem and were told not to board it as engineers tried to fix it. While the aircraft was en route to Jakarta, it had problems maintaining a constant altitude, with passengers stating that it was like a roller-coaster ride. The chief executive officer of Lion Air, Edward Sirait, confirmed that the aircraft had a technical issue on Sunday night, but this had been dealt with in accordance with maintenance manuals issued by the manufacturer. Engineers had declared that the aircraft was ready for takeoff on the morning of the accident.

A later report claimed that a third pilot was on the flight to Jakarta and told the crew to cut power to the stabiliser trim motors which fixed the problem. This method was a standard memory item in the 737 checklist. Subsequently, the National Transportation Safety Committee confirmed the presence of an off-duty Boeing 737 MAX 8 qualified pilot in the cockpit but did not confirm the role of the pilot in fixing the problem, and denied that there was any recording of the previous flight in the CVR of Lion Air Flight 610.

There was early speculation that the pitot tubes, used in the airspeed indication system, may have played a role in the crash, as they had contributed to previous similar incidents. Police Hospital Chief Musyafak said that an examination of the body parts indicated that it was unlikely that there had been an explosion or fire on board the aircraft.

Aviation experts noted that there were some abnormalities in the altitude and the airspeed of Flight 610. Just three minutes into the flight, the captain asked the controller for permission to return to the airport as there were flight control problems. About eight minutes into the flight, data transmitted automatically by the aircraft showed it had descended to about 5,000 feet, but its altitude continued to fluctuate. The mean value of the airspeed data transmitted by Flight 610 was around 300 knots, which was considered by experts to be unusual, as typically aircraft at altitudes lower than 10,000 feet are restricted to an airspeed of 250 knots. Ten minutes into the flight, the data recorded the aircraft dropping by more than 3,000 feet. The last recorded altitude of the aircraft was 2,500 feet.

On 5 November, the NTSC announced that Flight 610 was still intact when it crashed into the sea at high speed, citing the relatively small size of the pieces of debris. The impact was so powerful that the most substantial part of the aircraft was obliterated. The NTSC also stated that the engines of Flight 610 were still running when it crashed into the sea, indicated by the high RPM. Further examination of the aircraft's instruments revealed that one of the aircraft's airspeed indicators had malfunctioned for its last four flights, including the flight to Denpasar.

Two days later, on 7 November, the NTSC confirmed that there had been problems with Flight 610's angle of attack sensors. Thinking that it would fix the problem, the engineers in Bali then replaced one of them. Still the problem persisted on the penultimate flight from Denpasar to Jakarta. Just minutes after takeoff, the aircraft abruptly dived. The crew of that flight, however, had managed to control the aircraft and decided to fly at a lower than normal altitude. They then

The Altitude (in blue) and the Speed (in orange)graphs for the brief duration of the Lion Air flight 610. As can be seen, both varied considerably.

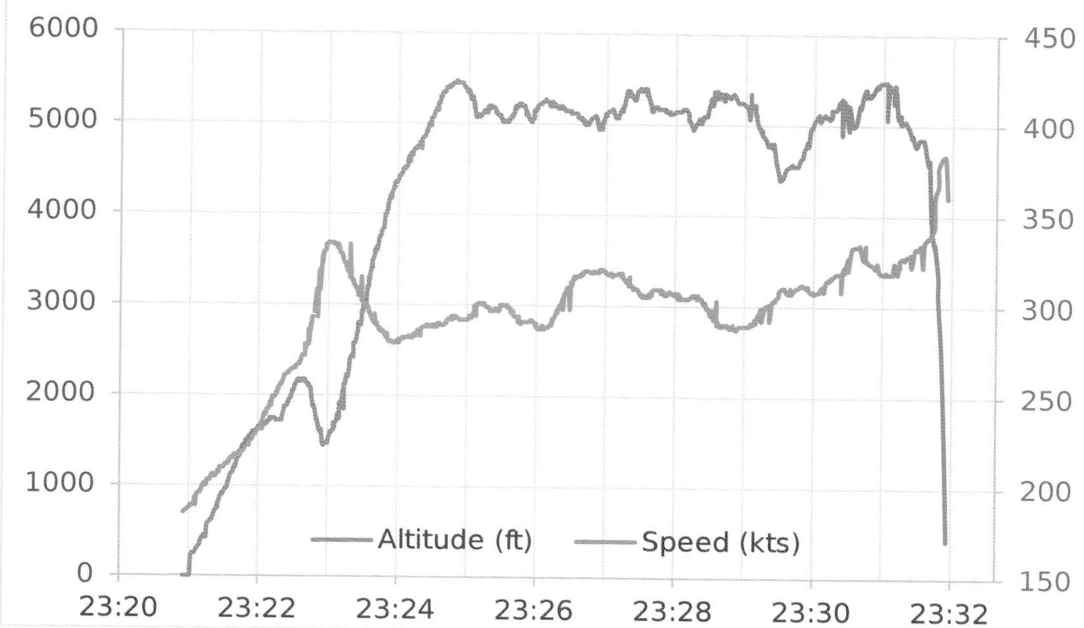

managed to land the aircraft safely and recorded a twenty-degree difference between the readings of the left AoA sensor and the right sensor. NTSC chief Soerjanto Tjahjono told the media that future reporting or actions, enacted to prevent similar problems on similar aircraft, would be decided by Boeing and U.S. aviation authorities.

On 28 November, Indonesian investigators said the Lion Air jet was not airworthy on the flight before the crash. Several relatives of the crash victims had already filed lawsuits against Boeing.

The same day the Indonesian National Transportation Safety Committee released its preliminary accident investigation report. After airspeed and altitude problems, an AoA sensor was replaced and tested two days earlier on the accident aircraft. False airspeed indications were still present on the subsequent flight on 28 October, which experienced automatic nose-down trim. The runaway stabiliser non-normal checklist was run, the electric stabiliser trim was turned off, and the flight continued with manual trim; the issues were reported after landing. Shortly after takeoff on 29 October, issues involving altitude and airspeed continued due to erroneous Angle of Attack data and commanded automatic nose-down trim via the Maneuvering Characteristics Augmentation System (MCAS). The flight crew repeatedly commanded nose-up trim over the final ten minutes of the flight. The preliminary report did not state whether the runaway stabiliser trim procedure was run or whether the electric stabiliser trim switches were cut out on the accident flight.

Boeing pointed to the successful troubleshooting conducted on 28 October as evidence that the MCAS did not change runaway stabiliser procedures, and emphasised the longstanding existence of procedures to cancel MCAS nose-down commands.

Ethiopian Airlines Flight 302

This was a scheduled international passenger flight from Addis Ababa Bole International Airport in Ethiopia to Jomo Kenyatta International Airport in Nairobi, Kenya. On 10 March 2019, the Boeing 737 MAX 8 aircraft which operated the flight crashed near the town of Bishoftu six minutes after takeoff, killing all 157 people aboard.

The aircraft, registered ET-AVJ with construction number 62450 and manufacturer's serial number 7243, was powered by two CFM International LEAP engines. The aircraft was manufactured in October 2018 and delivered on 15 November, making it around four months old at the time of the accident.

Flight 302 was, at the time the deadliest accident involving an Ethiopian Airlines aircraft, surpassing the fatal hijacking of Flight 961 resulting in a crash near the Comoros islands in 1996. At the time it was also the deadliest aircraft accident to occur in Ethiopia, surpassing the crash of an Ethiopian Air Force Antonov in 1982, which killed 73.

The conclusion of the Final Report into the crash of PK-LQP was published in October 2019.

III. CONCLUSION

The information obtained by the KNKT during its investigation of the Flight JT610 Accident demonstrates that the primary cause of the Flight JT610 Accident was Boeing's flawed design and development of MCAS. This conclusion is further supported by the independent analyses of Boeing's certification of the 737-8 MAX aircraft conducted by JATR and the NTSB.

Flight 302 was a scheduled international passenger flight from Addis Ababa to Nairobi. The aircraft took off from Addis Ababa at 08:38 local time with one hundred and forty-nine passengers and eight crew on board. One minute into the flight, the first officer, acting on the instructions of the captain, reported a flight control problem to the control tower. Two minutes into the flight, the plane's MCAS system activated, pitching the airliner into a dive toward the ground. The pilots struggled to control it and managed to prevent the nose from diving further, but the airliner continued to lose altitude. The MCAS then activated again, dropping the nose even further down. The pilots then flipped switches to disable the electrical trim tab system, which also disabled the MCAS software. However, in shutting off the electrical trim system, they also shut off their ability to trim the stabiliser into a neutral position with the electrical switch located on their yokes. The only other possible way to move the stabiliser would be by cranking the wheel by hand, but because the stabiliser was located opposite to the elevator, strong aerodynamic forces were pushing on it. As the pilots had inadvertently left the engines on full takeoff power, which caused the plane to accelerate at high speed, there was further pressure on the stabiliser. The pilots' attempts to manually crank the stabiliser back into position failed. Three minutes into the flight, with the aircraft continuing to lose altitude and accelerating beyond its safety limits, the captain instructed the first officer to request permission from air traffic control to return to the airport. Permission was granted, and the air traffic controllers diverted other approaching flights. Following instructions from air traffic control, they turned the aircraft to the east, and it rolled to the right. The right wing came to point down as the turn steepened. At 8:43, having struggled to keep the plane's nose from diving further by manually pulling the yoke, the captain asked the first officer to help him, and turned the electrical trim tab system back on in the hope that it would allow him to put the stabiliser back into neutral trim. However, in turning the trim system back on, he also reactivated the MCAS system, which pushed the nose further down. The captain and first officer attempted to raise the nose by manually pulling their yokes, but the aircraft continued to plunge toward the ground.

The aircraft disappeared from radar

Ethiopian Airlines 737 MAX 8 ET-AVL, the sister ship to the ill-fated ET-AVJ which crashed on 10 March 2019. *(Trymore Simango Collection)*

The fields of carnage that was the aftermath of the crash of ET-AVJ. *(both Trymore Simango Collection)*

screens and crashed at almost 08:44, six minutes after takeoff. Flight tracking data showed that the aircraft's altitude and rate of climb and descent were fluctuating. Several witnesses stated the plane trailed white smoke and made strange noises before crashing. The aircraft impacted the ground at nearly 700 mph. There were no survivors.

It crashed in the district of Gimbichu, Oromia Region, in a farm field near the town of Bishoftu, thirty-nine miles southeast of Bole International Airport. The impact created a crater about ninety feet wide and one hundred and twenty feet long, and the wreckage was driven up to thirty feet deep into the soil. Wreckage was strewn around the field, along with personal effects and body parts.

Shortly after the crash, police and a firefighting crew from a nearby Ethiopian Air Force base arrived and extinguished the fires caused by the crash. Police cordoned off the site, and Ethiopian Red Cross personnel and air crash investigators moved in. Together with local villagers, they sifted through the wreckage, recovering pieces of the aircraft, personal effects, and human remains. Trucks and excavators were brought in to assist in clearing the crash site. Human remains found were bagged and taken to Bole International Airport for storage in refrigeration units normally used to store roses destined for export before being taken to St. Paul's Hospital in Addis Ababa for storage pending identification. Personnel from Interpol and Blake Emergency Services, a private British disaster response firm contracted by the Ethiopian government, arrived to gather human tissue for DNA testing, and an Israel Police forensics team also arrived to assist in identifying the remains of the two Israeli victims of the crash. The Chinese railway construction firm CRSG, later joined by another construction firm, CCCC, brought in all large scale equipment including excavators and trucks. They recovered both black boxes on 11 March, with the first being found at 9:00 and the second flight recorder found at 13:00. The black boxes were given to Ethiopian Airlines and were sent to Paris for inspection by the Bureau d'Enquêtes et d'Analyses pour la sécurité de l'Aviation civile (BEA), the French aviation accident investigation agency.

The airline stated that the flight's 149 passengers had 35 different nationalities. The captain of the plane was 29-year-old Yared Getachew, who had been flying with the airline for almost nine years and had logged a total of 8,122 flight hours, including 1,417 hours on the Boeing 737. He had been a Boeing 737-800 captain since November 2017, and Boeing 737MAX since July 2018. At the time of the accident, he was the youngest captain in the airline. The first officer, 25-year-old Ahmed Nur Mohammod Nur, was a recent graduate from the airline's academy, with 361 flight hours logged, including 207 hours on the Boeing 737.

Ethiopian Prime Minister Abiy Ahmed offered his condolences to the families of the victims, while Ethiopian Airlines CEO Tewolde Gebremariam visited the accident site, confirmed that there were no survivors and expressed sympathy and condolences. Boeing issued a statement of condolence.

Ethiopian Airlines CEO Tewolde Gebremariam

The Ethiopian parliament declared 11 March as a day of national mourning. During the opening of the fourth United Nations Environment Assembly in Nairobi, a minute of silence was observed in sympathy for the victims. President Muhammadu Buhari of Nigeria, in his condolence message on behalf of the government and the people of Nigeria, extended his sincere condolences to Prime Minister Abiy Ahmed of Ethiopia, the people of Ethiopia, Kenya, Canada, China and all other nations who lost citizens in the crash.

On 11 March, the FAA commented that the Boeing 737 MAX-8 model was airworthy. However, due to concerns on the operation of the aircraft, they had ordered Boeing to implement design changes, effective by April. This directive stated that Boeing 'plans to update training requirements and flight crew manuals in response to the design change' to the aircraft's Maneuvering Characteristics Augmentation System (MCAS). The changes would also include enhancements to the activation of the MCAS and the angle of attack signal. Boeing stated that the upgrade was developed in response to the Lion Air crash but did not link it to the Ethiopian Airlines crash.

The Ethiopian Civil Aviation Authority, the agency responsible for studying civil aviation accidents in Ethiopia, started investigating the crash, and Boeing stated that it was prepared to work with the US National Transportation Safety Board and to assist Ethiopian Airlines. The United States FAA also became involved in the investigation.

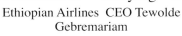

On 13 March 2019, the FAA announced that new evidence found on the crash site and satellite

data on Flight 302 suggested that the aircraft might have suffered from the same problem as Lion Air Flight 610 had. Investigators discovered the jackscrew that controlled the pitch angle of the horizontal stabiliser of Flight 302 was in the full 'nose down' position. The finding suggested that, at the time of the crash, Flight 302 was configured to dive, similar to Lion Air Flight 610. Due to this finding, some experts in Indonesia suggested that the Indonesian National Transportation Safety Committee (NTSC) should cooperate with Flight 302's investigation team. Later in the evening, the NTSC offered assistance to Flight 302's investigation team, stating that the committee and the Indonesian Transportation Ministry would send investigators and representatives from the government to assist with the investigation of the crash.

On 17 March, the Ethiopia's transport minister Dagmawit Moges announced that '...the black box has been found in a good condition that enabled us to extract almost all the data inside'. There was also confirmation that the preliminary data retrieved from the flight data recorder showed a clear similarity with those of Lion Air Flight 610.

In March 2020 Ethiopian crash investigators tentatively concluded that the crash of the Boeing 737 MAX was caused by design flaws. The conclusions, which said little or nothing about the performance of Ethiopian Airlines or its flight crew, raised concern with some participants in the investigation. The Ethiopian draft contrasted with conclusions by Indonesia's National Transportation Safety Committee after a prior 737 Max crash in October 2018. Rather than release a full report, the Ethiopian Aircraft Accident Investigation Bureau published an interim update one day before the anniversary of the 10 March 2019 crash.

Ethiopian investigators said a flawed flight control system triggered by faulty sensor data was at least partly to blame for the 2019 crash of the 737 MAX operated by Ethiopian Airlines.

Authorities in Ethiopia also said training on the MAX airliners provided by Boeing 'was found to be inadequate' adding that the flight control system, known as MCAS, was activated four times as pilots struggled mightily to regain control of the plane before the crash.

The conclusions, which included recommendations, were only in draft form and could be altered before release, said the people, who requested anonymity to discuss the sensitive matter. It was possible the US National Transportation Safety Board could request changes to the report or offer a dissenting opinion.

Chapter Thirteen

Grounding - and Ways Out

The Ethiopian Airlines crash on 10 March 2019, so soon after the Lion Air event, both involving the 737 MAX, saw worldwide media attention dramatically focus on the aircraft. Events started to move rapidly and were catastrophic for the airliner.

11 March - China: The Civil Aviation Administration of China ordered all domestic airlines to suspend operations of all 737 MAX 8 aircraft by 18:00 local time pending the results of the investigation, thus grounding all 96 Boeing 737 MAX airliners in China.

USA: The FAA issued an affirmation of the continued airworthiness of the 737 MAX. The FAA stated that it had no evidence from the crashes to justify regulatory action against the aircraft.

Indonesia: Nine hours after China's grounding, the Indonesian Ministry of Transportation issued a temporary suspension on the operation of all eleven 737 MAX-8 aircraft in Indonesia. A nationwide inspection on the type was expected to take place on 12 March to 'ensure that aircraft operating in Indonesia are in an airworthy condition'.

Mongolia: Civil Aviation Authority of Mongolia (MCAA) said in a statement 'MCAA has temporarily stopped the 737 MAX flight operated by MIAT Mongolian Airlines from 11 March 2019'.

12 March - Singapore: the Civil Aviation Authority of Singapore, 'temporarily suspended operation of all variants of the 737 MAX aircraft into and out of Singapore.

India: Directorate General of Civil Aviation (DGCA) released a statement 'DGCA has taken the decision to ground the 737 MAX aircraft immediately, pursuant to new inspections'.

Turkey: Turkish Civil Aviation Authority suspended flights of 737 MAX 8 and 9 type aircraft being operated by Turkish companies in Turkey, and stated that they were also reviewing the possibility of closing the country's airspace for the same.

South Korea: Ministry of Land, Infrastructure and Transport (MOLIT) advised Eastar Jet, the only airline of South Korea to possess Boeing 737 MAX aircraft, to ground their models, and three days later issued a NOTAM (Notice to Airmen) to block all Boeing 737 MAX models from landing and departing from all domestic airports.

Europe: The European Union Aviation Safety Agency (EASA) suspended all flight operations of 737 MAX-8 and 737 MAX-9 in Europe. In addition, EASA published a Safety Directive, published at 18:23, effective as of 19:00 UTC, suspending all commercial flights performed by third-country operators into, within or out of the European Union of the above mentioned models.

Canada: Minister of Transport Marc Garneau said it was premature to consider groundings and that, 'If I had to fly somewhere on that type of aircraft today, I would.'

Australia: The Civil Aviation Safety Authority banned Boeing 737 MAX from Australian airspace.

Malaysia: The Civil Aviation Authority of Malaysia suspended the operations of the Boeing 737 MAX 8 aircraft flying to or from Malaysia and transiting in Malaysia.

13 March - Canada: Minister of Transport Marc Garneau, prompted by receipt of new information, said 'There can't be any MAX-8 or MAX-9 flying into, out of or across Canada', effectively grounding all 737 MAX aircraft in Canadian airspace.

USA: President Donald Trump announced that United States authorities would ground all 737 MAX-8 and MAX-9 aircraft in the United States.

After the President's announcement, the FAA officially ordered the grounding of all 737MAX-8 and 9 operated by US airlines or in the United States airspace. The FAA did allow airlines to make ferry flights without passengers or flight attendants in order to reposition the aircraft in central locations.

Hong Kong: The Civil Aviation Department banned the operation of all 737 MAX aircraft into, out of and over Hong Kong.

Panama: The Civil Aviation Authority grounded its aircraft.

Vietnam: The Civil Aviation Authority of Vietnam banned Boeing 737 MAX aircraft from flying over Vietnam.

New Zealand: The Civil Aviation Authority of New Zealand suspended Boeing 737 MAX aircraft from its airspace.

Mexico: Mexico's civil aviation authority suspended flights by Boeing 737 MAX-8 and MAX-9 aircraft in and out of the country.

Brazil: The National Civil Aviation Agency suspended the 737 MAX-8 aircraft from flying.

Summer in Europe saw a number of airlines put their aircraft in light-hearted schemes. Here Norwegian airline Braathens 737-505 LN-BRX disconnects from the tug after pushback from the stand at London Gatwick in 1993. Alongside each artwork is the name and age of the child who created it. *(Earl Gray collection)*

Another wild and wacky scheme - probably created in during a break in one of Amsterdam's famous coffeeshops - is this design applied to PH-HSI, a 737-8K2(WL) of Transavia. The 'Peter Pan vakantieclub' or Peter Pan Holiday Club Foundation has been organizing trips for young people with chronic illness since 1996. During that trip they do not have to think about their illness, but they can make new friends and have a great time together. All members of the Peter Pan Holiday Club Foundation are employees of Transavia. A permanent team of volunteers organizes four trips per year, to Lapland, Fuerteventura, Barcelona and Mallorca. Over the years, more than a thousand young people have made an unforgettable journey with Peter Pan. *(author)*

Colombia: Colombia's civil aviation authority banned Boeing 737 MAX-8 planes from flying over its airspace.

Chile: The Directorate General of Civil Aviation banned Boeing 737 MAX-8 flights in the country's airspace.

Trinidad and Tobago: The Director General of Civil Aviation banned Boeing 737MAX-8 and -9 planes from use in civil aviation operations within and over Trinidad and Tobago.

14 *March - Taiwan:* The Civil Aeronautics Administration banned Boeing 737 MAX from entering, leaving or flying over Taiwan.

Japan: Japan's transport ministry banned flights by Boeing 737 MAX 8 and 9 aircraft from its airspace.

16 *March - Argentina:* The National Civil Aviation Administration (ANAC) closed airspace to Boeing 737MAX flights.

27 *June - Belgium:* issued a NOTAM extending the 737-8 MAX and -9 MAX ban until 2020.

At the time of its grounding, the MAX was operating 8,600 flights per week. About thirty Boeing 737 MAX aircraft were airborne in US airspace when the FAA grounding order was announced. The airliners were allowed to continue to their destinations and were then grounded. In Europe, a number of flights were diverted when grounding orders were issued; for example, an Israel-bound Norwegian 737 MAX returned to Stockholm, and two Turkish Airlines MAXs flying to Britain, one to Gatwick Airport and the other to Birmingham, turned around and flew back to Turkey.

On 11 June Norwegian Flight DY8922 attempted a ferry flight from Málaga, Spain to Stockholm, Sweden. Such flights could only be flown by pilots meeting certain European Aviation Safety Agency (EASA) qualifications, and with no other cabin crew or passengers. The flight plan contained specific parameters to avoid MCAS intervention, flying at lower altitude than normal with flaps extended, and autopilot on. However, the aircraft was refused entry into German airspace, and diverted to Châlons Vatry, in France.

In a rare exemption, Transport Canada approved eleven flights in August and September 2019, partly to maintain the qualifications of senior Air Canada training pilots, because the airline had no earlier-generation 737s within its fleet. The airline used the MAX during planned maintenance movements, and ultimately flew it to Pinal Airpark in Arizona for storage.

In early October 2019, Icelandair moved two of its five MAX 8s for winter storage in the milder climate of northern Spain, making the entire flight with flaps extended to prevent MCAS activation.

Focus on the FAA - and Criminal Proceedings.

Buried in the early part of the timeline of events, on 11 March 2019, a US federal grand jury issued a subpoena on behalf of the US Justice Department for documents related to the development of the 737 MAX. On 19 March the US Department of Transportation requested the Office of Inspector General to conduct an audit on the 737 MAX certification process. The FBI then joined the criminal investigation into the certification. FBI agents reportedly visited the homes of Boeing employees in 'knock-and-talks'.

President Trump's executive order on 13 March to ground all Boeing 737 MAX 8s was a necessary step. However, many thought that it should have been taken directly by the FAA - that it came from the White House instead spoke volumes to a profound crisis of public confidence in the agency.

The roots of this crisis are found in a major change the agency instituted in its regulatory responsibility in 2005. Rather than naming and supervising its own 'designated airworthiness representatives' the agency allowed Boeing - and other manufacturers who qualified under the revised procedures - to select specified employees to certify the safety of their aircraft. This laissez-faire certification system would save the aviation industry nearly $25 million between 2006 and 2015, the FAA said at the time. This was a pittance when compared with Boeing's $81 billion in revenue for 2012. It seems to be no coincidence that the committee that helped develop this process was made up of industry members. Essentially, aircraft makers had persuaded the FAA to let them certify their own airliners so they could save money.

UR-SQE, a 737-75C(W) of SkyUp Airlines, in the striking livery of Football Club Shakhtar Donetsk, a Ukrainian professional football club from the city of Donetsk. In 2014, due to the armed conflict in eastern Ukraine, the club was forced to move to Lviv, and has since early 2017 played in the city of Kharkiv whilst having its office headquarters and training facilities in Kyiv. *(Dee Diddley Collection)*

US President Donald J Trump and Boeing CEO Dennis Muilenburg.

Then, in 2013 a scandal emerged - the certification of the Boeing 787's lithium-ion battery, and brought into question the cosy relationship between manufacturer and regulator. Following the grounding of the 787 Dreamliner in January this relationship contributed to the astonishing swiftness with which the aircraft was approved to return to commercial flight.

The FAA let Boeing help write the safety standards, develop the testing protocol and then perform those tests. In 2008, a year after standards for the battery system were approved with special conditions on the containment and venting of the batteries, stricter industry guidelines for these batteries were released. But the FAA did not require the 787 to meet those new guidelines.

Boeing initially estimated that there was the potential for one battery failure incident in ten million flight hours. As it turned out, smoke and fire broke out in batteries on two separate 787s in just the first 52,000 flight hours. Even Boeing's chief engineer on the 787, Mike Sinnett, acknowledged to the NTSB that one of the battery tests had been inadequate and was not 'conservative enough'.

Lawyers, analysts and industry experts criticized CEO Muilenberg's delivery of Boeing's public statements as contradictory and unconvincing. They said Boeing refused to answer tough questions and accept responsibility, defended the airliner's design and certification while 'promising to fix the plane's software', delayed to ground aircraft and issue an apology, and yet was quick to assign blame towards pilot error.

Boeing issued a statement after each crash, saying it was ' deeply saddened' by the loss of life and offered its 'heartfelt sympathies to the families and loved ones' of the passengers and crews. Boeing said it was helping with the Lion Air investigation and was also sending a technical team to assist in the Ethiopia investigation.

As non-US countries and airlines began grounding the 737 MAX, Boeing stated: 'at this point, based on the information available, we do not have any basis to issue new guidance to operators, saying that in light of the Ethiopian Airlines crash, the company would privatize the roll-out ceremony for the first completed Boeing 777X.

The MAX grounding concerned the general public and worldwide regulators; consequently, the harsh spotlight of concern started to focus on the FAA, who began to re-evaluate its certification process. The FAA was seeking consensus with other regulators to approve the return to service to avoid suspicion of undue cooperation with Boeing. The International Air Transport Association (IATA) had also made a similar statement calling for more coordination and consensus with training and return to service requirements.

The FAA certifies the design of aircraft and components that are used in civil aviation operations. It is performance-based, proactive, centred on managing risk, and focused on continuous improvement. As with any other FAA certification, the MAX certification included reviews to show that system designs and the MAX complied with FAA regulations; ground tests and flight tests; evaluation of the airliner's required maintenance and operational suitability;

Western Pacific Airlines - also known as WestPac - was known for using their planes as billboards. The airline started in Colorado Springs in 1995 with eight Boeing 737-300 planes under the name Commercial Air. Soon after launch, Oklahoma millionaire and owner of Gaylord Hotels founder Edward Gaylord invested in the carrier, in the process requesting the carrier to switch names from Commercial Air to Western Pacific. The airline quickly gained popularity through its unique aircraft. WestPac operated an all Boeing 737-300 fleet, but most were used as logojets to help promote businesses that cooperated with WestPac or were part of the airline's route map. Some of these advertisements included Thrifty Rental Car, the Stardust and Sam's Town Casinos in Las Vegas, and Fox TV's hit show 'The Simpsons.' *(Matt Black collection)*

collaboration with other civil aviation authorities on aircraft approval.The FAA denied claims that they had authorised aircraft makers to police themselves or self-certify their aircraft. However, the FAA delegated to Boeing vast portions of the certification process as per the FAA and Industry Guide to Product Certification. This Guide described how to plan, manage, and document an effective and efficient product certification process and working relationship between the FAA and the companies. This process was intended to result in the more effective use of FAA and Applicant resources with a systems approach to oversight focused on risk-based areas.

Therefore, the FAA had latitude in delegating responsibilities to Boeing. The practical implementation of the accountability framework was for the FAA to exercise its discretion on the level of involvement necessary to make a finding that the applicant had shown compliance with all the applicable requirements before issuing a design approval. The FAA's responsibilities, starting with top management, included:

a. Enabling Applicants to maximise delegation within their projects, processes, and procedures (Organization Designation Authorizations and non-Organization Designation Authorisations known as ODAs and non ODAs).

b. Applying risk-based oversight processes, behaviours and tools within Applicant projects, processes, and procedures. Efficient FAA Level of Project Involvement (LOPI) will be determined by a risk-based oversight model based on observed compliance capability of the applicant with a systems approach that incorporates an audit process by the FAA and Holder after the completion of the project.

c. Enable the applicant's path towards the Applicant Showing Only (ASO), state. In this role, the FAA accepts the applicant's compliance data as compliant without FAA or designee review when the applicant's capability has been determined competent by the FAA.

During design and construction of the MAX, the FAA delegated a large number of safety assessments to Boeing itself, a practice that had been standard for years, but several FAA insiders believed that the recent level of the delegation went too far.

Boeing had 1,500 people in its ODA, under supervision of a FAA team of forty-five people, of which only twenty-four were engineers. By 2018, the FAA was letting Boeing certify ninety-six per cent of its work. With the installation of MCAS, initially, as a new system or device on the amended type certificate, the FAA retained the oversight of this equipment. However, the FAA later released it to the Boeing ODA based on comfort level and thorough examination, Daniel Elwell of the FAA said in March. 'We were able to assure that the ODA members at Boeing had the expertise and the knowledge of the system to continue going forward'.

In March 2019, reports emerged that Boeing performed the original System Safety Analysis, and FAA technical staff felt that managers pressured them to sign off on it. Boeing managers also pressured engineers to limit safety testing during the analysis. It came to light that in a 2016 survey Boeing found almost forty per cent of 523 employees working in safety certification felt 'potential undue pressure' from managers.

In August 2019, reports of friction between Boeing Co. and international air-safety authorities emerged. A Boeing briefing was stopped short by the FAA, EASA, and other regulators, who complained that Boeing had 'failed to provide technical details and answer specific questions about modifications in the operation of MAX flight-control computers'.

A few days later, on 22 August the FAA announced that it would invite pilots from around the world, intended to be a representative cross-section of 'ordinary' 737 pilots, to participate in simulator tests as part of the recertification process, at a date to be determined. The evaluation group sessions were to be one of the final steps in the validation of flight-control computer software updates and would involve around thirty pilots, including some first officers

OO-LTU, a 737-33A of Virgin Express. This airline started on 23 April 1996, when the Virgin Group (with chairman Richard Branson) bought the Belgian leisure airline EBA - EuroBelgian Airlines - founded by Victor Hasson and Georges Gutelman and was rebranded as Virgin Express. It also took over EBA's fleet of 737s and operated this type of aircraft from there. *(Bob Lehat Collection)*

Luxair's 737-5C9 LX-LGO seen at London Heathrow was delivered new back in December 1992. It passed through a whole procession of operators. *(Bob Lehat Collection)*

with multi-crew pilot licences which emphasised simulator experience rather than flight hours. The FAA hoped that the feedback from pilots with more varied experience would enable it to determine more effective training standards for the aircraft.

When the FAA grounded the MAX aircraft on 13 March, Boeing stated it 'continues to have full confidence in the safety of the 737 MAX'. However, after consultation with the FAA, the NTSB, and aviation authorities and its customers around the world, Boeing determined – out of an abundance of caution and in order to reassure the flying public of the aircraft's safety – to recommend to the FAA 'the temporary suspension of operations of the entire global fleet of 737 MAX aircraft.'

After the grounding, Boeing suspended 737 MAX deliveries to customers, but continued production at a rate of fifty-two aircraft per month. In mid-April, the production rate was reduced to forty-two aircraft per month.

Fixing it in the software.

Details as to the operation of the MCAS is covered in a later chapter, but for continuity of events, mention must be made of it here. Between the Ethiopian accident and US groundings, Boeing stated that upgrades to the MCAS flight control software, cockpit displays, operation manuals and crew training were underway due to findings from the Lion Air crash. Boeing anticipated software deployment before April and said the upgrade would be made mandatory by an FAA Airworthiness Directive.

On 14 March Boeing reiterated that pilots could always use manual trim control to override software commands and that both its Flight Crew Operations Manual and 6 November 2018 bulletin offered detailed procedures for handling incorrect angle-of-attack readings.

The FAA stated it anticipated clearing the software update by 25 March, allowing Boeing to distribute it to the grounded fleets, but on 1 April announced the software upgrade was delayed because more work was necessary.

Daniel Elwell of the FAA.

Three days later, Boeing CEO Dennis Muilenburg acknowledged that MCAS played a role in both crashes. His comments came in response to the public release of preliminary results of the Ethiopian Airlines accident investigation, which suggested that the pilots had indeed performed the recovery procedure. Muilenburg stated it was 'apparent that in both flights' MCAS activated due to 'erroneous angle of attack information'. He said the MCAS software update and additional training and information for pilots would 'eliminate the possibility of unintended MCAS activation and prevent an MCAS-related accident from ever happening again'. Boeing reported that ninety-six test flights were flown with the updated software.

Dennis A Muilenburg (*b*. 1964) started work at Boeing as an intern in 1985. He held numerous management and engineering positions on a whole series of Boeing programmes, including the X-32, Boeing's entry in the Joint Strike Fighter competition; Boeing's participation in the Lockheed Martin F-22 Raptor fighter; the YAL-1 747 Airborne Laser; the High Speed Civil Transport; and the Condor unmanned reconnaissance aircraft. He later served as vice president of the Boeing Combat Systems division. Muilenburg served as president and chief executive officer of Boeing Integrated Defense Systems (later renamed Boeing Defense, Space & Security (BDS) from September 2009 till 2015.

JATR Chair Christopher A Hart.

In December 2013, Muilenburg became the president of Boeing. In June 2015, Boeing announced that Muilenburg would succeed James McNerney as CEO, who was stepping down after ten years in that role. He became CEO in July 2015. In February 2016, it was announced that Muilenburg would succeed McNerney as chairman.

In an earnings call that took place on 24 April, Muilenburg again said the aircraft was properly designed and certificated and denied that any 'technical slip or gap' existed. He said there were 'actions or actions not taken that contributed to the final outcome'. On 29 April, he claimed that the pilots did not 'completely' follow the procedures that Boeing had outlined. He said Boeing was working to make the airplane even safer.

By 4 August Boeing reported that they conducted around 500 test flights with updated software.

On 18 September 2019, FAA administrator Stephen Dickson, who had succeeded Daniel Elwell, said that he would not certify the MAX until he flew the aircraft himself (Dickson had previously been a pilot for Delta Air Lines and operated the Boeing 737). He said he would fly the airliner using the new software following

the certification flight.

Then, on 2 October, 2019 it came out that Boeing had convinced FAA regulators to relax certification requirements in 2014 - an act that reduced the development cost of the MAX. The revelation came about because according to a number of current and former FAA officials, instead of increasing its oversight powers, the FAA had been pressing ahead with plans to reduce its hands-on oversight of aviation safety further.

On 11 December, 2019, Dickson announced that MAX would not be recertified before 2020, and reiterated that FAA did not have a timeline. The following day, Dickson met with Boeing chief executive Dennis Muilenburg to discuss Boeing's unrealistic timeline and the FAA's concerns that Boeing's public statements might be perceived as attempting to force the FAA into quicker action.

Hearings in Congress

Meanwhile, there were things happening in other areas. In March 2019, the US Congress announced an investigation into the FAA approval process for the 737 MAX. Members of Congress and government investigators expressed concern about FAA rules that allowed Boeing to extensively 'self-certify' aircraft. FAA acting Administrator Daniel Elwell said 'We do not allow self-certification of any kind'.

Meanwhile, on 2 April after receiving reports from whistleblowers regarding the training of FAA inspectors who reviewed the 737 MAX type certificate, the Senate Commerce Committee launched a second Congressional investigation focusing on FAA training of the inspectors.

The FAA had provided misleading statements to Congress about the training of its inspectors, most possibly the same people who oversaw the MAX certification, according to the findings of an Office of Special Counsel (OSC) investigation released in September. The OSC was also tasked with investigating reports from whistleblowers (also written as whistle-blower or whistleblower); a person who exposes secret information or activity that is deemed illegal, unethical, or not correct within a private or public organisation. Its report inferred that FAA safety inspectors 'assigned to the 737 MAX had not met qualification standard'.

Southwest Airlines' 737-8H4(W) ETOPS N8302F seen taxiing out at Manchester - Boston Regional Airport, New Hampshire. *(Matt Black Collection)*

Left: Chairman of the House Transportation and Infrastructure Committee Peter DeFazio.

Right: Secretary, US Department of Transportation, The Hon. Elaine L Chao

The OSC sided with the whistleblower, pointing out that internal FAA reviews had reached the same conclusion. In a letter to President Donald Trump, the OSC found that sixteen of the twenty-two FAA pilots conducting safety reviews, some of them assigned to the MAX two years ago, 'lacked proper training and accreditation'.

Safety inspectors participated in Flight Standardization Boards, which ensured pilot competency by developing training and experience requirements. FAA policy required both formal classroom training and on-the-job training for safety inspectors.

Special Counsel Henry J. Kerner wrote in the letter to the President, 'This information specifically concerns the 737 MAX and casts serious doubt on the FAA's public statements regarding the competency of agency inspectors who approved pilot qualifications for this aircraft'.

In September, Daniel Elwell of the FAA disputed the conclusions of the OSC, which found that aviation safety inspectors assigned to the 737 MAX certifications did not meet training requirements.

This, however, did not mean that the FAA were not prepared to investigate further. On 19 April a Boeing 737 MAX Flight Control System Joint Authorities Technical Review (JATR) team was commissioned by the FAA to investigate how it approved MCAS, whether changes needed to be made in the FAA's regulatory process and whether the design of MCAS complied with regulations. On 1 June Ali Bahrami, FAA Associate Administrator for Aviation Safety, chartered the JATR to include representatives from FAA, NASA and the nine civil aviation authorities of Australia, Brazil, Canada, China, Europe (EASA), Indonesia, Japan, Singapore and UAE.

On 27 September the JATR chair, Christopher A Hart, said that FAA's process for certifying new aircraft was not broken but needed improvements rather than a complete overhaul of the entire system. He added 'This will be the safest airplane out there by the time it has to go through all the hoops and hurdles'.

The JATR said that FAA's limited involvement and inadequate awareness of the automated MCAS safety system resulted in an inability of the FAA to provide an independent assessment. The panel report added that Boeing staff performing the certification were also subject to undue pressures which further eroded the level of assurance in this system of delegation.

The JATR team identified a number of areas related to the evolution of the design of the MCAS where the certification deliverables had not been updated during the certification programme to reflect the changes to this function within the flight control system. In addition, the design assumptions were not adequately reviewed, updated, or validated; possible flight deck effects were not evaluated; the SSA and functional hazard assessment (FHA) were not consistently updated; and potential crew workload effects resulting from MCAS design changes were not identified. Nor had Boeing carried out a thorough verification by stress-testing of the MCAS.

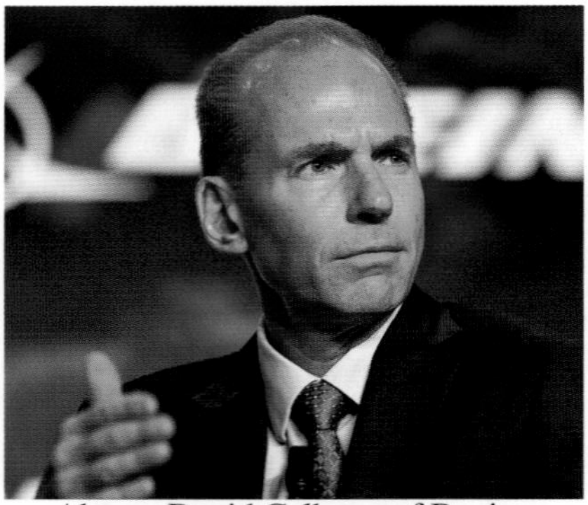

Above: David Calhoun of Boeing

Below: Boeing Board member Admiral Edmund Giambastiani (Retd).

The JATR found that Boeing exerted what they termed as 'undue pressures' on the Boeing Organization Designation Authorization engineering unit members, who had FAA authority to approve design changes.

On 7 June Peter DeFazio (D; Or), chairman of the House Transportation and Infrastructure Committee, wrote to The Honorable Elaine L Chao, Secretary US Department of Transportation and Acting FAA Administrator Daniel K Elwell: 'Nine weeks ago this week we wrote to the FAA regarding the tragic deaths of 346 people, including eight Americans, in two separate crashes of Boeing 737 MAX aircraft. The letter informed Acting Administrator Elwell that our Committee was investigating these accidents and related issues regarding the certification of the 737 MAX aircraft and it requested numerous records to help inform our investigation. Since then our respective staff have had continuous, weekly, ongoing discussions about these records requests. In addition, several weeks ago Secretary Chao committed to us that the Department would be forthcoming and cooperative with the Committee regarding our investigation into the 737 MAX aircraft'.

'The Committee has been patient and respectful of the time and resources available to the FAA to abide by this request, and accommodating in seeking to narrow the scope of this request at the behest of the FAA. In addition to these weekly discussions, more than four weeks ago, Committee staff agreed to a bipartisan meeting at FAA Headquarters to further discuss these records requests in an additional accommodation to the FAA. Until yesterday, the only records the Committee had received from the FAA were lists of documents Committee staff agreed to review and prioritize in an effort to further accommodate the FAA's requests and one technical document that is public and widely available on the Internet. Yesterday, we received a second delivery of documents that was composed of a total of eighty-two pages of records. Although we appreciate that these documents were finally provided, we are concerned about the slow pace of the FAA's response. The FAA informed our staff weeks ago that they had already identified

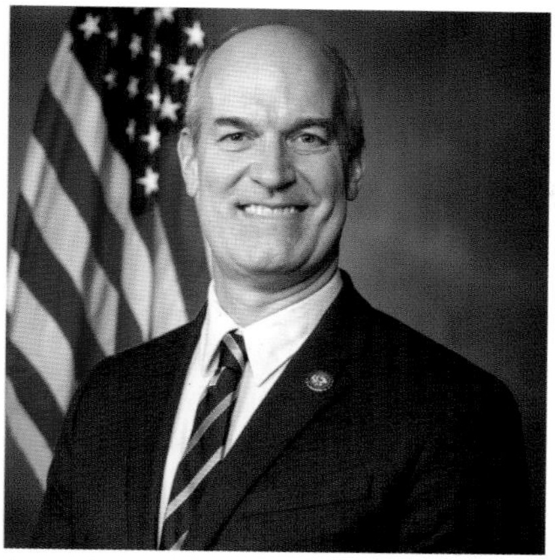

Left: Leader of the House Transportation and Infrastructure Committee Aviation Sub-Panel Rick Larsen

Below: Vice-President and Chief Engineer Boeing Commercial Airplanes Division John Hamilton.

hundreds of thousands of responsive e-mails and tens of thousands of responsive documents'.

'To say we are disappointed and a bit bewildered at the ongoing delays to appropriately respond to our records requests would be an understatement. We are concerned about the FAA's lack of a sufficient response to our request. In addition to a written explanation regarding the continual delay in providing the Committee, which has clear and unambiguous oversight authority over the FAA, with these records, please provide us with a detailed timeline regarding the delivery of the documents we originally asked for in our April 1st letter'.

'The Lion Air and Ethiopian Airlines accidents have focused public attention on the FAA. As leaders of the Committee on Transportation and Infrastructure, we have an obligation to the public to oversee and improve US transportation infrastructure, including the safety of the aviation industry. You both also have an obligation to the Committee to respond to the investigation we have launched regarding the recent 737 MAX accidents. We trust that the lack of an effective response to the Committee so far will be quickly resolved to allow the Committee's investigation to proceed, which we hope will help to identify issues that will improve aircraft safety, the certification process and oversight of these issues'.

'We look forward to a timely response to this letter. Thank you for your prompt attention to this matter'

The Subcommittee on Aviation met on 19 June to hold a hearing called the *Status*

SX-BBU was a 737-33A, of Aegean Airlines, later to be fitted with winglets and operated as G-GDFB for jet2. *(Kirk Smeeton Collection)*

of the Boeing 737 MAX: Stakeholder Perspectives. The hearing was intended to gather views and perspectives from aviation stakeholders regarding the Lion Air Flight 610 and Ethiopian Airlines Flight 302 accidents, the resulting international grounding of the Boeing 737 MAX aircraft, and actions needed to ensure the safety of the aircraft before returning them to service.

On 17 July, a number of representatives of crash victims' families, in testimony to the House Transportation and Infrastructure Committee - Aviation Subcommittee, called on regulators to re-certificate the MAX as a completely new aircraft. They also called for wider reforms to the certification process and asked the committee to grant protective subpoenas so that whistle-blowers could testify even if they had agreed to a gag order as a condition of a settlement with Boeing.

In a senate hearing on 31 July, the FAA defended its administrative actions following the Lion Air accident, noting that standard protocol in ongoing crash investigations limited the information that could be provided in the airworthiness directive.

In September, a Congressional panel asked Boeing's CEO to make several employees available for interviews, to complement the documents and the senior management perspective already provided. That month, Boeing's board called for changes to improve safety. Peter DeFazio said Boeing declined his invitation to testify at a House hearing. 'Next time, it won't just be an invitation, if necessary'.

Inside Boeing, Muilenburg was carrying out a board-ordered revamp that gave directors an unfiltered view of concerns flagged by employees, while adding initiatives of his own to sharpen the focus on safety.

'Dennis has, and has always had, the full, unequivocal and enthusiastic support of the Boeing board of directors,' the company's lead outside director, David Calhoun, said in a statement. Muilenburg has been 'incredibly supportive' of the board's new Aerospace Safety Committee, said Admiral Edmund Giambastiani, a Boeing director who was to lead the oversight panel. 'Dennis

A close-up study of TUI's OO-MAX *Tenerife Alegria,* a 737 Max-8, showing in detail the LEAP-1B engines and the light-emitting diode (LED) landing lights buried in the wing root. *(Kirk Smeeton collection)*

has the board's confidence,' he said in an interview 25 September.

Admiral Giambastiani had been elected to its board of directors on 8 October 2009. *The Seattle Times* reported that 'In a statement, Boeing chairman and CEO Jim McNerney indicated that the addition of Giambastiani, who was the second-highest ranking officer in the US military, is intended to boost Boeing's influence with the Pentagon.'

FAA Administrator The Hon. Stephen M Dickson.

'Boeing has taken to heart the aerospace industry's tradition of studying tragedies for ways to make flying safer', Muilenburg said in the interview at Boeing's Chicago headquarters. Reading that at the time, one had to wonder how long Muilenburg would remain at his desk.

That same month the House Transportation and Infrastructure Committee announced that Dennis Muilenburg would testify before Congress accompanied by John Hamilton, chief engineer of Boeing's Commercial Airplanes division, and Jennifer Henderson, 737 chief pilot. In October 2019, the House asked Boeing to allow a employee who filed an internal ethics complaint to be interviewed.

On 29 October Muilenburg and Hamilton appeared at the House hearing under the title *Aviation Safety and the Future of Boeing's 737MAX,* which was the first time that Boeing executives addressed Congress about the MAX accidents. The hearing came on the heels of the removal of Dennis Muilenburg's title as chairman of the Boeing board a week before and was intended to examine issues associated with the design, development, certification, and operation of the 737MAX following two accidents. The committee first heard from Boeing on actions taken to improve safety and the company's interaction with relevant federal regulators. The second panel was comprised of government officials and aviation experts discussing the status of Boeing 737MAX and relevant safety recommendations.

The next day the House made public an internal email discussion between Boeing employees raising concerns about MCAS design in the exact scenario blamed for the two crashes: 'Are we vulnerable to single Angle of Attack sensor failures with the MCAS implementation?' Committee members discussed another internal document, stating that a reaction longer than ten seconds to an MCAS malfunction '...found the failure to be catastrophic.' The hearing's key revelation was insider knowledge of vulnerabilities amid a hectic rate of production.

After the testimony of Boeing's CEO, Peter DeFazio, Rick Larsen, leader of its aviation sub-panel, wrote a letter to other lawmakers on 4 November updating them as to 'progress'. With the letter were a lot of headings and bullet-points. They do not make for flowing reading, but are important as to their content.

'Last week at the Transportation and Infrastructure Committee's fourth hearing on the Boeing 737 MAX, we received testimony from Boeing CEO and President Dennis Muilenburg, who was accompanied by Vice President and Chief Engineer of Boeing's Commercial Airplanes division, John Hamilton. This hearing was a key step in our Committee's investigation into the design, development, and certification of the

Boeing 737 MAX, an investigation that began three days after the second deadly crash involving this particular airplane model.

'At our hearing, the Committee revealed some of the new information, emails, and records that our investigation has uncovered over the last seven months:

MCAS Design

- A preliminary design of Boeing's 737 MAX included an MCAS alert in the cockpit but was later removed.
- The actual operation of MCAS on the two deadly flights violated Boeing's own design criteria for MCAS which required that, 'MCAS shall not interfere with dive recovery,' and 'MCAS shall not have any objectionable interaction with the piloting of the airplane.'
- Boeing officials knew that if it took a pilot more than 10 seconds to react to erroneous MCAS activation, the result could be 'catastrophic.'
- In December 2015, a Boeing engineer in the division that designed MCAS asked, 'Are we vulnerable to single [angle-of-attack] sensor failures with the MCAS implementation?'

Pilot Training:

- Beginning in 2013, Boeing leadership had a clear plan to ensure the MAX did not require 'simulator training' for pilots that is expensive and time-consuming for airline customers. In 2014, Boeing marketed the MAX to airlines as a '[n]o simulator required' plane, years before the Federal Aviation Administration made a decision on what kind of pilot training was required.
- Boeing officials told the FAA at least twice that MCAS didn't need to be in the Flight Crew Operating Manual or training and asked to have it removed from these training manuals because they claimed MCAS only operated, 'way outside of the normal operating envelope.' However, MCAS activated within the normal operating envelope on Lion Air Flight 610 and Ethiopian Airlines Flight 302.

Undue Pressure

- A November 2016 internal Boeing employee survey found 39 percent of respondents perceived undue pressure and 29 percent were concerned about the consequences of reporting undue pressure.
- A former Boeing supervisor working on the 737 MAX final assembly line raised serious safety issues with senior Boeing management in June 2018, four months before the Lion Air crash. The employee - who wished to remain anonymous - specifically raised the issues of 'schedule pressure' and wrote in an email: 'Frankly right now all my internal warning bells are going off. And for the first time in my life, I'm sorry to say that I'm hesitant about putting my family on a Boeing airplane.' The employee wrote that he was so concerned that he recommended shutting down the production line. 'I don't make this recommendation lightly,' he wrote. 'I know this would take a lot of planning, but the alternative of rushing the build is far riskier. Nothing we do is so important that it is worth hurting someone.' A few months after he left Boeing, the employee wrote to Boeing CEO Dennis Muilenburg after the Lion Air crash about his concerns.

Chair of the House Committee on Transportation and Infrastructure Peter DeFazio and Chair of the Subcommittee on Aviation Rick Larsen both issued statements: 'Our hearing last week was an important step in our investigation, but it certainly did not mark the end. Based on what we heard from Mr. Muilenburg and Mr. Hamilton in front of our Committee, we have a litany of new questions for both Boeing and the FAA about the failures that led to the tragic and unnecessary deaths of 346 innocent people'.

'To summarize our key concerns, our investigation shows that from almost the start, Boeing had a bad design on MCAS with a single point of failure. Then, Boeing couldn't even meet its own design requirements. MCAS was fundamentally flawed, and according to Boeing's own analysis, could result in catastrophic consequences in certain cases. What's more, Mr. Muilenburg's answers to our questions were consistent with a culture of concealment and opaqueness and reflected the immense pressure exerted on Boeing employees during the development and production of the 737 MAX. Boeing leadership has said that if company officials knew during the design of the MAX what they know now about some of the technical flaws and other issues, they would have done things differently. Our investigation has already shown that Boeing leadership was aware of many of the problems that engineers are now attempting to fix during the design and development phase of the 737 MAX'.

'We were surprised by Mr. Muilenburg's apparent lack of awareness of rather critical decisions being made within his own company, including something that concerns us deeply, which is Boeing's attempt to move legal proceedings related to the MAX overseas and out of the US court system'.

'The bottom line is that there are a lot of unanswered questions, and our investigation has a long way to go to get the answers everyone deserves, especially the families of the crash victims. We are grateful that during the hearing, many Members stopped to recognize the family members who were in attendance. That small gesture is important because we can't lose sight of the human toll of the mistakes that were made on the Boeing 737 MAX. The victims' loved ones deserve a thorough investigation from our Committee about how the regulatory system and the law failed, and that's exactly what our Committee intends to do'.

'In the coming days and weeks, our Committee will push ahead on our investigation and we will keep you updated on the next hearing'.

Peter DeFazio and Rick Larsen then wrote to The Honorable Stephen M Dickson, Administrator FAA on 7 November, questioning the way the FAA made decisions on two safety-related items with potentially catastrophic consequences: rudder cable protection on the Boeing 737

The scene at Boeing Field, Seattle that Boeing wished they could hide, and everyone wanted to photograph. Twenty-two 737 MAX airliners, with covers on the tyres and tape around all the openings to keep the notorious Washington State rain out. *(J Duford)*

MAX and lightning protection for the Boeing 787 Dreamliner. The document was complex and technical, but is interesting in recording the facts regarding what had been happening 'behind the scenes'.

'As you know, our Committee has been investigating the design and development of Boeing's 737 MAX, the Federal Aviation Administration's (FAA) certification processes, and related issues. While our investigation is ongoing, we are concerned about two additional safety issues about which we have received detailed information. Both appear to involve serious, potentially catastrophic safety concerns raised by FAA technical specialists that FAA management ultimately overruled after Boeing objected. These incidents raise questions about how the agency weighs the validity of safety issues raised by its own experts compared to objections raised by the aircraft manufacturers the FAA is supposed to oversee'.

'Boeing 737 MAX Rudder Cable Protection from Uncontained Engine Failure. The first issue involves the adequacy of rudder cable protection on the Boeing 737 MAX from an uncontained engine failure and the possibility of severance of the cable and a potentially catastrophic loss of control'.

'In 2014, a manager in the FAA's Transport Airplane Directorate issued a memo to a higher official in the FAA's Aircraft Certification Service asserting that Boeing had not incorporated adequate protection into the 737 MAX rudder cable as required by 14 CFR 25.903(d) *[FAA Transport Airplane Directorate memo to FAA Aircraft Certification Service, 3/ 10/14, updated 9/22/14, p. 1.]*. ' The memo noted Boeing's previous agreement to show compliance with the latest guidance, found in Advisory Circular 20-128A, which applied lessons learned from the 1989 United Airlines Flight 232 accident near Sioux City, Iowa, in which debris from an uncontained engine failure severed hydraulic lines, resulting in a crash landing that left 112 people dead. Boeing objected to making changes to the design of the 737 MAX rudder cable, arguing that changes would be impractical and noting the company's concern about the potential impact on 'resources and program schedules.'

The FAA's Transport Airplane Directorate found Boeing's position unacceptable and stated its intention to release an issue paper to Boeing 'requiring they protect the rudder cable from [uncontained engine failure] per AC 20-128A.' *[FAA Transport Airplane Directorate memo to FAA Aircraft Certification Service, 3/10/14, updated 9/22/14, p. 2.]*'

'In 2015, the FAA drafted an issue paper, finalized in 2016, that offered Boeing a chance to establish compliance without implementing a design change. *[Issue Paper: Engine Rotor Burst and Rudder Mechanical Flight Control Cables, 7 / 24/ 15.]* At least six FAA specialists refused to concur. Strangely, the issue paper also suggested that, based on the 'excellent' service history of the different engine on the prior version of the 737, the FAA 'expected' the new, larger LEAP engine would have a similarly low rate of uncontained engine failures.' From an analytical perspective, that argument appears to be nonsensical since the FAA was making an unfounded conclusion about the reliability of a then-unproven new engine based on the performance of a completely different older engine. This statement, however, was not part of a showing or finding of compliance'.

'When concern about the issue paper was submitted to the FAA's safety review process, a panel was established to review the matter. On January 13, 2017, the panel recommended that the FAA '[i]nform Boeing there is currently insufficient information, data and coordination between the FAA and Boeing such that a determination of compliance can be made'.

'The panel also rejected Boeing's position that design changes were impractical, finding, instead, that two design changes were, in fact, practical.' The panel also made clear the inappropriateness of consideration of reliability of a previously approved engine to demonstrate compliance, and that the new LEAP engine was sufficiently different from its predecessor that past performance of the older engine would not be relevant in predicting the new engine's performance.' *[SRP Item 10 SME Panel - Findings and Recommendations to the SRP Safety Oversight Board, 1/13/17.]* Despite these concerns, the 737 MAX gained certification from the FAA two months later in March 2017'.

'It is our understanding that non-concurrence by FAA technical specialists is fairly infrequent and not to be taken lightly. In addition, my staff has been told that it was virtually unprecedented for six or more FAA specialists to jointly non-concur on a single issue, highlighting the gravity of their concerns regarding the rudder cable issue. Despite all of this, in June 2017, the FAA's Transport Airplane Directorate upheld the controversial issue paper. *[FAA Transport Airplane Directorate memo to FAA Aircraft Certification Service, 6/30/17]*'.

DeFazio and Larsen then moved on to the topic of lightning protection for Boeing 787 fuel tanks: 'Our Committee has also received information and documents suggesting Boeing implemented a design change on its 787 Dreamliner lightning protection features to which multiple FAA specialists ultimately objected. In addition to the merits of the safety risks the FAA experts raised, it is also of great concern that Boeing reportedly produced approximately 40 airplanes prior to the FAA's approval of the design change. If accurate, that is an astonishing fact that suggests either willful neglect of the Federal aviation regulatory structure or an oversight system in need of desperate repair'.

'The change involves the removal of copper foil from zone 3 of the wing of the 787 Dreamliner, which could result in significantly higher conducted currents in that zone as well as increase the number of

corendon Airlines (incorporated as Turistik Hava Taşımacılık A.Ş.) is a Turkish leisure airline headquartered and based at Antalya Airport. The airline was founded in 2004 with flight operations starting in April 2005. The Dutch sister company Corendon Dutch Airlines was founded in 2011 and the Maltese sister company Corendon Airlines Europe was founded in 2017. Here TC-MKS, their 737 MAX-8, is seen with cascade thrust reversers deployed while landing at Amsterdam. *(Phil McCrackin Collection)*.

ignition sources in the fuel tanks'.

'Boeing's design change failed to comply with Special Condition 25-414-SC, which requires Boeing to show that a fuel tank ignition would be extremely improbable'.

'Lightning strikes on aircraft are a fairly routine occurrence. This is true of the 787 Dreamliner, an aircraft built of more than 50 per cent carbon fiber composites. 'While incredibly lightweight and strong, such aircraft composites are not inherently conductive, thus requiring additional protective coatings to mitigate lightning strike damage,' according to a technical blog post on lightning protection measures.' [reference Jennifer Segui, *'Protecting Aircraft Composites from Lightning Strike Damage,' CO MSOL Blog, June 11, 2015]* Two years ago, a British Airways Boeing 787 was struck by lightning shortly after it departed London's Heathrow airport. When the aircraft landed in Chennai, India, it was discovered the aircraft had more than 40 holes in the fuselage from the lightning strike. *'Boeing 787 Grounded for a Week after Lightning Strike,' August 5, 2017, Air Insight'* Three years earlier, in October 2014, a United Airlines Boeing 787 was struck by lightning leaving London's Heathrow airport en route to Houston, Texas. *'Brett Macdonald, 'Why Superior Lightning Strike Protection Is Vital In Aerospace,' June 14, 2018, Dexmet Corporation'.*

Above: Ryanair EI-DWM, a 737-8AS(W), on final approach to London Stansted.

Left: The interior of a Ryanair aircraft, complete with advertisements on the overhead bins and safetly briefing on the seatbacks. *(both Dee Didley)*

HL8223, a 737-9B5ER(W) of Korean Air awaits its turn to get airborne. *(Phil McCrackin Collection)*.

'On February 22, 2019, the FAA's Boeing Aviation Safety Oversight Office (BASOO) formally rejected Boeing's lightning protection design change. Apparently, Boeing appealed the decision, and a meeting was held on February 27, 2019, during which a Boeing official reportedly stated that Boeing employees had discussed the issue with the FAA's Associate Administrator for Aviation Safety. On March 1, 2019, FAA management reversed course, and accepted Boeing's position.'

'It is our understanding that the FAA has recently tasked Boeing with performing a numerical risk assessment of the fuel tank explosion risk from lightening related ignition sources that addresses each Model 787 configuration that is determined to exist to date. The FAA apparently plans to use this assessment 'to determine if any corrective actions to reduce the risk of a fuel tank explosion should be required by airworthiness directive action.' [reference Oct. 15, 2019 letter from FAA Seattle Aircraft Certification Office Branch to Boeing Organization Designation Authorization]'.

'While we appreciate that the FAA is finally taking some action on this issue, we are deeply concerned that the agency is just now asking Boeing to provide analysis to enable the FAA 'to determine if any corrective actions' are required. It appears Boeing took actions that may have violated FAA requirements in the first place by taking unilateral steps to change the design of the aircraft's lightning protection system. Asking Boeing to now review its own work in the aftermath of those events, if true, to help the FAA determine what corrective actions Boeing may need to take seems woefully inadequate to ensure the safety of the flying public. In addition, this process will take several months, and we would like to know how the FAA is satisfied that the risk is sufficiently low that these 787s can continue flying in revenue service before the numerical risk assessment is completed'.

'The two cases above regarding the 737 MAX and the 787 Dreamliner suggest that the opinions and expert advice of the FAA's safety and technical experts are being circumvented or sidelined while the interests of Boeing are being elevated by FAA senior management. There may be reasonable explanations for FAA management overriding the decisions of its own technical experts at the behest of the manufacturer it

Gear up! SX-BMC, a 737-42J of Olympic Airways gets airborne from Geneva.
(Phil McCrackin Collection).

regulates, but we would like a clear accounting of those explanations in the two instances described above. We respectfully request that you please provide:

1. A detailed explanation of how the FAA decided to overrule its own safety specialists with regard to the two safety issues described above, including the process FAA relied upon to make those determinations and who at FAA made those ultimate decisions. In addition, please describe what the FAA is doing to ensure that these two issues do not pose a risk to the flying public.
2. An explanation of what the FAA is doing to ensure that manufacturers do not have an incentive to attempt end-runs around FAA technical specialists by going to senior FAA management.
3. A list of all lightning protection-related regulations, requirements, or standards applicable to the 787 aircraft certification at the time Boeing produced such aircraft before FAA-approval of the Boeing design change, and a description of FAA actions taken in response to any deviations of those regulations, requirements, or standards by Boeing.
4) An explanation of the FAA's conclusion that the 787s produced in response to the design change are safe to operate in revenue service before Boeing completes its numerical risk assessment of the overall fuel tank explosion risk from lightning related ignition sources, and before the FAA has had an opportunity to evaluate that assessment.

Please respond to this request by November 21, 2019'.

On 11 December, during a hearing of the House Committee on Transportation titled *The Boeing 737MAX: Examining the Federal Aviation Administration's Oversight of the Aircraft's Certification,* held at 2167 Rayburn House Office Building, an internal

FAA review dated 3 December 2018, was released, which predicted a high MAX accident rate over the type's lifetime using the TARAM - Transport Airplane Risk Assessment Methodology - system if it kept flying with MCAS unchanged. It predicted fifteen fatal crashes of the roughly 4,800 MAX in service over the next 45 years, if the MAX kept flying with MCAS unchanged. Boeing also did a similar TARAM risk analysis for the MAX after the Lion Air crash, with results matching those of the FAA. The FAA had shared its TARAM analysis with Boeing. It seemed that the FAA assumed that the emergency airworthiness directive sufficed until Boeing delivered a fix.

Massachusetts Institute of Technology Professor Arnold Barnett refuted the report based on the loss of two aircraft out of only 400 delivered. He said there would be twenty-four crashes per year for a fleet size of 4,800, thus the FAA underestimated the risk by a factor of 72.

The next day, FAA Administrator told US lawmakers it knew the risks of further accidents following the Lion Air crash involving a Boeing 737 MAX but did not issue a grounding order until after a second incident.

Committee chair Peter DeFazio said in his opening statement: 'You can be sure this committee will continue to be aggressive in our oversight efforts to determine what went so horribly wrong and why, and we will not rest until we have enacted legislation to prevent future unairworthy airplanes from slipping through the regulatory cracks and into airline service.'

Over a question whether a mistake was made in this regard, the FAA's chief, Stephen Dickson, responded, 'Obviously the result was not satisfactory', and in response to repeated questions he admitted the agency had made a mistake at some point in the process. He said the report found an unacceptable level of risk, but the FAA started to take action with the emergency order.

Peter DeFazio said that the committee's investigation '...has uncovered a broken safety culture within Boeing and an FAA that was unknowing, unable, or unwilling to step up, regulate and provide appropriate oversight of Boeing'.

DeFazio accused the FAA of 'rolling the dice on the safety of the travelling public' by letting the Max continue to fly following the Lion Air crash. He went on to say that a single FAA manager overruled the 'unanimous judgement of more than a dozen FAA safety experts'.

The committee also revealed that it had become aware of an instance in which Boeing placed pressure on the FAA to overrule engineers' concerns on safety-critical issues. A whistleblower said engineers had determined that an uncontained engine failure on the Max could send shrapnel through the rudder control cables, meaning pilots could likely lose control of the aircraft during the initial climb off the runway or during a take-off roll.

Chair of the Subcommittee on Aviation Rick Larson later said 'the FAA must fix its credibility problem'.

The statements came after Boeing CEO Dennis Muilenburg was told by lawmakers they believed the manufacturer had put profit before safety on the 737 MAX programme, highlighting concerns that the company had engaged in '...a pattern of deliberate concealment' during the certification process.

Re-certification

The new aircraft certification process is entirely different from what the FAA had in the past. Over the decades, the American regulator only certified the overall type of an airliner. It used to delegate other routine day-to-day activities, such as certifying individual aircraft of an already accredited type - the planemaker.

In November 2019, the FAA announced that it had withdrawn Boeing's authority, previously held under the Organization Designation Authorization, to issue airworthiness certificates for individual new 737 MAX aircraft. After the overall grounding is lifted, the FAA will issue such certificates directly; aircraft already delivered to customers will not be affected. In the same month, the FAA pushed back at Boeing's attempts to publicise a certification date, saying the agency would take all the time it needed.

In a letter sent to Boeing, the FAA announced its changed certification approach for every 737 MAX aircraft, stating that 'the FAA retain authority to issue airworthiness certificates and export certificates of airworthiness for all 737 MAX airplanes.'

The move by the FAA reflected its attempt to regain goodwill and show its independency from Boeing. Notably, the American regulator had faced criticism for having close ties with the planemaker. Also, it had been blamed for delegating critical MAX certification processes to the planemaker employees.

So it was that the FAA publicly announced it would take all the time it needed to verify the safety of MAX. The regulator also made it clear that it had not completed the review of design changes and associated pilot training.

By all accounts, the company had been working to fix the software issue in MAX's flight-control system which some software engineer-sources called a kludge, regarded as mainly being what caused the two deadly crashes. Boeing needed the FAA and other global regulators' certification for the updated software. However, the FAA's latest comment further cast a shadow over Boeing MAX's return to service. A 'kludge' is computer technical-speak for a poorly programmed piece of software, a piece of hardware cobbled together from spare parts, or a project plan created by someone with no real experience with the task at hand.

Then, on Christmas Eve, a new batch of Boeing internal documents related to the 737 MAX surfaced, painting what was termed 'a very disturbing picture regarding employees' concerns about safety', a House of Representatives committee reported.

The latest documents were sent over late Monday, the same day Boeing CEO Dennis Muilenburg resigned, effective immediately, after a string of troubling disclosures. Both the FAA and the House Transportation and Infrastructure Committee, which had held a series of hearings about the 737 MAX, acknowledged receiving the Boeing documents but did not disclose the contents, saying they were under review. However, it was reported in the *Seattle Times* that the documents included more internal communications involving former Boeing 737 chief test pilot Mark Forkner, who, in records previously disclosed, described problems in the development of the flight control system blamed in the two crashes. His 2016 missive to a colleague also talked of basically, but unknowingly, lying to regulators.

As early as 2016 Mark Forkner, then Boeing's chief technical pilot for the 737 MAX during its certification, expressed misgivings about a feature since implicated in two fatal crashes, calling its handling performance 'egregious' in a series of electronic messages to another Boeing employee.

The FAA reported that Boeing alerted the Transportation Department of instant messages between the pilot and another employee of the planemaker in October 2019; the regulator said that Boeing had been aware of the communications for months, something that the regulator was not happy with: 'The FAA finds the substance of the document concerning. The FAA is also disappointed that Boeing did not bring this document to our attention immediately upon its discovery.'

The November 2016 instant messages were exchanges between Mark Forkner, and another 737 technical pilot, Patrik Gustavsson.

The entire 2016 text-message exchange - as they were sent - which was released to the House Committee on Transportation reads as follows:

Forkner: dude, log off!

Gustavsson: You too!!!

I just logged on to check my schedule. I have so much to do that I want to work from home I just cant get stuff done in the office

Forkner: nah, I'm locked in my hotel room with an ice cold grey goose, I'll probably fire off a few dozen inappropriate emails before I call it a night

Gustavsson: LMAO!!!!

Forkner: this job is insane

Gustavsson: So did you get anything done in the sim today? Or what is the normal chaos there?

Forkner: although it must be easy compared to working as a tech pilot for RYR [Ryanair]

Gustavsson: Well it's different here. We are pretty busy here for sure.

Forkner: actually this one is pretty stable, and I signed off some DRs, but there are still some real fundamental issues that they claim they're aware of

Gustavsson: What I hated about Ryanair was the extreme pressure they put on people Ok, that's good

Forkner: so I just need to start being a dick to make you quit?

Gustavsson: LOL, that's it!

Forkner: alright, no more mr nice guy! actually I'd cry uncontrollably if you left

I'd ask for a job in sales where I can just get paid to drink with customers and lie about how awesome our airplanes are

Gustavsson: I'd cry if anyone in our group left.

Forkner: Oh shocker alerT!

MCAS is now active down to M.2 It's running rampant in the sim on me at least that's what Vince thinks is happening

Gustavsson: Oh great, that means we have to update the speed trim descritption in vol 2

Forkner: so I basically lied to the regulators (unknowingly)

Gustavsson: it wasnt a lie, no one told us that was the case

Forkner: I'm levelling off at like 4000 ft, 230 knots and the plane is trimming itself like craxy I'm like, WHAT?

Gustavsson: that's what i saw on sim one, but on approach

I think thats wrong

Left: Mark Forkner.

Right: Patrik Gustavsson.

Forkner: granted, I suck at flying, but even this was egregious

Gustavsson: No, i think we need aero to confirm what its supposed to be doing

Forkner: Vince is going to get me some spreadsheet table that shows when it's supposed to kick in. why are we just now hearing about this?

Gustavsson: I don't know, the test pilots have kept us out of the loop

It's really only christine that is trying to work with us, but she has been too busy

Forkner: they're all so damn busy, and getting pressure from the program

Gustavsson: That is true, I wouldnt want to be them

Ok, its time to log off

Forkner: ok later man

Gustavsson: I'll work from home tomorrow, be online all day later'

As can be seen, Forkner and Gustavsson raised multiple concerns about the automated flight control system, including not being given data by the company's test pilots and Forkner describing his alarm at simulator tests in which he encountered troubling behaviour in the system. Boeing had earlier assured the aviation regulator that the MCAS was benign and didn't need to be included in the aircraft's flight manuals, according to one person familiar with the issue and the FAA had approved the company's request.

The messages were just one more turn in the saga of the 737MAX. While the planemaker repeatedly said that it followed proper procedures to certify the airliner, the communications show that senior pilots at the company were concerned about a critical aspect of its design and were worried that regulators had been misled.

FAA Administrator Steve Dickson sent a terse letter to CEO Muilenburg on 18 October. 'Last night I reviewed a concerning document that Boeing provided late yesterday to the Department of Transportation. I understand that Boeing discovered the document in its files months ago. I expect your explanation immediately regarding the content of this document and Boeing's delay in disclosing the document to its safety regulator.'

Boeing had given the text messages to the Justice Department in February 2019 - a month before the second 737MAX crash - but did not inform the FAA because the exchange was sensitive to a criminal investigation then being conducted. Knowing it would need to disclose the document to the House Transportation and Infrastructure Committee, which was

JY-SOB is a 737-33V of Fly Jordan, was originally G-EZYS of easyJet. Fly Jordan, formerly known as Solitaire Air, is a Jordanian charter carrier based at Queen Alia International Airport. *(Phil McCrackin Collection).*

conducting an investigation of the airliner's certification, Boeing finally shared the document the day before Dickson's letter with the Department of Transportation's general counsel.

'This is more evidence that Boeing misled pilots, government regulators and other aviation experts about the safety of the 737MAX,' Jon Weaks, president of the Southwest Airlines Pilots Association, said in a statement on 18 October. 'It is clear that the company's negligence and fraud put the flying public at risk.'

The House Transportation and Infrastructure Committee Chairman Peter DeFazio said the exchange 'essentially constitutes a smoking gun of the pressure exerted by Boeing executives to get the 737 MAX into service quickly'. DeFazio called on Muilenburg, who was scheduled to speak before the committee on 30 October, to resign.

The House Transportation and Infrastructure Committee's investigation into FAA's original certification of the 737 MAX had viewed hundreds of thousands of documents, including emails that DeFazio called disturbing that, in addition to interviews and whistle-blower reports, pointed to a pattern of '...massive production pressure exerted from on high on Boeing employees to get this plane out, and apparently it was 'get it out no matter what'.

In reply to all this, Boeing said it had been cooperating with the committee's investigation. 'As part of that cooperation, today we brought to the committee's attention a document containing statements by a former Boeing employee,' company spokesman Gordon Johndroe said in an email.. 'We will continue to cooperate with the Committee as it continues its investigation. And we will continue to follow the direction of the FAA and other global regulators, as we work to safely return the 737 MAX to service.'

A lawyer representing Boeing in the matter, McGuireWoods LLC's Richard Cullen, said in a statement that: 'The Boeing Company timely produced the Mark Forkner Instant Messenger document to the appropriate authorities.'

Other Whistleblowers

The FAA received at least four calls from potential Boeing employee whistleblowers about issues with the 737 MAX. The calls began coming in within hours of Ethiopian investigators releasing a preliminary report on the second of those crashes, that of Ethiopian Airlines Flight 302.

The four calls came in through a special hotline set up by the FAA for employees or the public to report problems. They were being evaluated by FAA investigators as part of ongoing probes into the 737 MAX and its

Air Canada's 737 MAX-8 C-FSCY was delivered to the airline on 4 December 2017, and is seen here at a wintry Toronto Pearson International Airport. Note the fleet number '502' on both the nosewheel door and at the tip of the vertical fin. *(Rick O'Shea Collection)*

certification.

One of the whistleblower complaints filed internally was from a Boeing engineer who helped design the MAX's flight controls.

In the document, later shared with US Justice Department investigators, Curtis Ewbank said his managers and Michael Teal, the 737 MAX's chief project engineer, repeatedly rejected adding the synthetic airspeed system, used on the 787 Dreamliner, and taking data from a number of sources which might have counteracted a sensor that malfunctioned in both crashes as a safety backstop on the basis of cost and impact on pilot training.

Ewbank worked on the 737 MAX flightdeck systems that pilots used to monitor and control the airliner. In his complaint to Boeing, he said managers were urged to study a backup system for calculating the aircraft's airspeed. Such equipment, Ewbank said, could detect when the angle-of-attack sensors, which measure the aircraft's attitude in the sky, were malfunctioning and preventing other systems from relying on that faulty information.

Curtis Ewbank noted in his complaint, 'It is not possible to say for certain that any actual implementation of synthetic airspeed on the 737 MAX would have prevented the accidents', but he said Boeing's actions on the issue pointed to a culture that emphasised profit, in some cases, at the expense of safety.

Throughout the development of the MAX, Boeing tried to avoid adding components that could force airlines to train pilots in flight simulators, something that would incur costs of tens of millions of dollars over the life of an aircraft. As already stated, significant changes to the MAX could also have required the more onerous approval process for a new airliner, rather than the streamlined certification for a derivative model.

According to Mr. Ewbank's complaint, Ray Craig, a chief test pilot of the 737, and other engineers wanted to study the possibility of adding the synthetic airspeed system to the MAX. But a Boeing executive decided not to look into the matter because of its potential cost and effect on training requirements for pilots. 'I was willing to stand up for safety and quality, but was unable to actually have an effect in those areas,' Ewbank said in the complaint. 'Boeing management was more concerned with cost and schedule than safety or quality.'

His account, and the description of the

With the worldwide grounding of the 737 MAX, it fell to other 737s, such as this 737-924(W) N32404 of United Airlines, to fill in where they could. *(Matt Black Collection)*

HL7510, a 757-48E of Air Busan, a low-cost airline based in Busanjin-gu, Busan, South Korea. It is a subsidiary of Asiana Airlines. *(Matt Black Collection)*

system's benefits on the 787 Dreamliner, were backed up by a former senior Boeing employee involved in the discussions, who spoke on the condition of anonymity because of the continuing Justice Department investigation. The former Boeing employee, who worked on the MAX, confirmed that executives had discussed the system. He said they had determined that trying to install such new technology on the 737 MAX, would be too complicated and risky for the project, which was on a tight schedule.

But the former Boeing employee said Curtis Ewbank's complaint overstated the importance of such a system and understated the complexity of adding it to the 737 MAX. This employee said that Boeing had installed the system only on the 787 Dreamliners, noting that it was unclear how or whether the MAX could similarly calculate synthetic airspeed because it had fewer sensors. The employee also did not recall Boeing executives citing the potential impact on pilot training when deciding not to study adding the system.

Company spokesman, Gordon Johndroe, said in a statement, 'Boeing offers its employees a number of channels for raising concerns and complaints and has rigorous processes in place, both to ensure that such complaints receive thorough consideration and to protect the confidentiality of employees who make them.'

The FAA confirmed that the 5 April calls were from current and former Boeing employees alleging possible issues related to the angle of attack sensor and the MCAS anti-stall system that relies on data from the sensor. One of the claims dealt with damage to the wiring of an AOA sensor from a foreign object.

Boeing said it could not verify the report but in a statement said '... Safety and quality are absolutely at the core of Boeing's values. Speaking up is a cornerstone of that safety culture and we look into all issues that are raised.'

A Boeing source was sceptical of that particular whistleblower allegation, saying that as far as they knew, there had been '...no reported issues...at all' with foreign object debris damage to angle of attack sensors or their wiring.

Not involving the 737, but certainly related, Boeing had issues with foreign object debris - known as FOD - being found in the company's 787 Dreamliner assembled at its South Carolina plant, including metal swarf discovered by the FAA in aircraft Boeing certified as debris-free as recently as 2017.

Twice in 2019, the US Air Force stopped accepting delivery of the 767-based KC-46 aerial refueller because FOD was found inside the newly delivered aircraft. The KC-46 is built in Everett, Washington.

Boeing's failure to deliver its new KC-

Australian air Express VH-XMR, a 737-376F VH-XMR leased from QANTAS is captured taxiing out at Brisbane. *(Matt Black collection)*

46 Pegasus on schedule could force the military to keep using ageing tankers to meet its needs, the US Transportation Command's top officer said. 'We've got to figure out a way to mitigate the delayed fielding of the KC-46,' General Stephen Lyons said on 28 January 2020, while speaking at the Atlantic Council in Washington DC.

'The Air Force has already planned on the retirement of a select number of its KC-135 Stratotankers and KC-10 Extenders, even as the Boeing-run KC-46 program struggles to develop and field new tankers. If we're not careful we're going to see a real dip, a bathtub, in taskable tails for the joint force,' Lyons said, referring to the number of mission-ready tankers.

In 2008, the service selected a modified tanker version of the Airbus A330 commercial airliner over Boeing's offer of a 767 derivative. But that decision was nullified after Boeing complained about the selection process, and in 2011 won the renewed tender for a tanker capable of in-flight refuelling.

However, Boeing failed to provide the tankers since winning the $44 billion project. The company had promised to deliver eighteen combat-ready aircraft by 2017, but Air Force leaders said that was unlikely to happen for several more years.

In the meantime, the Air Force has taken delivery of thirty tankers for training purposes, while Boeing continues to fix some of the aircraft's problems, including an issue with the cameras used to guide the extended 59-foot boom into a fuel receptacle on the receiving aircraft.

Some solutions to the tanker shortfall include the 'ability to retain some number of legacy tails,' Lyons said, referring to the KC-135s and KC-10s.

Another of the potential whistle-blower calls dealt with concerns over the shutoff switches for MCAS.

Previously the Senate Commerce Committee launched an investigation into the FAA certification process, citing whistleblower claims of improperly trained FAA inspectors working on the MAX. House Transportation and Infrastructure Committee investigators had reportedly been speaking with potential whistleblowers. At the time, it was unclear if any of these whistleblowers allegations overlapped.

The revelations raised questions about Boeing's Chief Executive Officer Dennis Muilenburg. They surfaced less than two

Sporting a WiFi 'bump' on top of the fuselage, VH-XZB, a 737-838(W) of QANTAS is caught landing at Melbourne. *(Matt Black collection)*

weeks before Muilenburg was scheduled to appear before lawmakers in Washington to answer questions on the airliner. Boeing directors stripped Muilenburg of his chairmanship on 11 October in the wake of a damaging report from a multinational review of the airliner's certification.

What's in a name?

Plenty of observers – among them US President Donald Trump – advised Boeing to rebrand the airliner before its eventual return to service. The US president tweeted in April: 'If I were Boeing, I would FIX the Boeing 737 MAX, add some additional great features, & REBRAND the plane with a new name.'

Boeing told reporters at the Paris Air Show, held over 17 to 23 June 2019 that it would consider changing the name to help the model return to the skies, although the company said it was not working on a name change in its most recent statement on the matter: 'We remain open-minded to all input from customers and other stakeholders but have no plans at this time to change the

Operating as Aussie 312, this Royal Australian Air Force 373-700BBJ, A36-001 is seen landing at Geneva. *(Phil McCrackin Collection)*.

Air Vanuatu is an airline with its head office in Air Vanuatu House, Port Vila, Vanuatu. It is Vanuatu's national flag carrier, operating to Australia, New Zealand and points in the South Pacific. Its main base is Bauerfield International Airport, Port Vila. YJ-AV1 is a 737-838(W). *(Matt Black collection)*

name of the 737 MAX.'

On the first day of the show - with much fanfare - Boeing and the International Airline Group issued a Press Release announcing the signing of a 'letter of intent' for two hundred 737 MAXs: 'International Airlines Group (IAG) has signed a letter of intent with Boeing for 200 B.737 aircraft to join its fleet. The LOI is subject to formal agreement.

The mix of 737-8 and 737-10 aircraft would be delivered between 2023 and 2027 and would be powered by CFM Leap engines. It is anticipated that the aircraft would be used by a number of the Group's airlines including Vueling, LEVEL - a low cost subsidiary - plus British Airways at London Gatwick airport. Willie Walsh, IAG chief executive, said: 'We're very pleased to sign this letter of intent with Boeing and are certain that these aircraft will be a great addition to IAG's short-haul fleet.

'We have every confidence in Boeing and expect that the aircraft will make a successful return to service in the coming months having received approval from the regulators'.

Read that again please. Notice something? Where is the word MAX? Rather than referring to the airliners as the 737 MAX-8 and 737 MAX-10, they referred to them as the 737-8 and 737-10.

In a 'Note to Editors', section attached to the release, IAG stated: 'The list price is approximately US$117 million for the Boeing 737-8 and US$131 million for the 737-10.

'The list price is the sum of the airframe list price, engine option list price and the price of optional features against which price concessions are made. IAG has negotiated a substantial discount from the list price'.

The deal was seen as a crucial vote of confidence in the troubled airliner and provided a significant boost to the US manufacturer after its rival Airbus appeared to be stealing a march on it with big orders and the launch of a competing model. Although more than 5,000 had been ordered, the IAG deal was the first sale of

Re-branding or not?

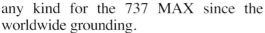

A pair of Ryanair 737 MAXs seen in storage at Renton, EI-HAW on the left with '737 MAX' on the nose and EI-HAY, below, with 737-8200 painted on it. While both designations are correct one has to question what was going on! *(BJ)*

any kind for the 737 MAX since the worldwide grounding.

So was the 'name change' just a slip of the typing-finger? Maybe, but just under a month later - on 15 July 2019 - photographs surfaced of the nose of a recently-completed 737 MAX registered EI-HAY for Ryanair and parked at Boeing's Renton facility – with the 'MAX' name no longer on the nose. The Irish airline had one hundred and thirty-five Boeing 737 MAX 200s on order. Five of these had already been produced, though due to the grounding they were not flying. The nose of the fourth plane built said '737 MAX'. Meanwhile, the nose of the fifth airliner built said '737-8200'. So it appears that Ryanair was adopting the IAG '737-8' branding, but were adding '200' to it to differentiate it from the other 737-8s. Several media reports appeared claiming that Ryanair was rebranding the 737 MAX, in the aftermath of the re-engined type's grounding.

Although the switch is notable, the name, contrary to a number of reports, was not a new designation for the aircraft, and actually predated both the MAX grounding and the two fatal accidents which led to the type's suspension from operations.

The European Union Aviation Safety Agency had been referring to the 737-8200 in documents such as its 2015 annual activity report, published in June 2016, almost a year before the first-ever 737 MAX delivery, and this designation has frequently been included in US federal regulatory filings from Boeing and the US FAA since at least early 2017.

EASA documents on seating capacity and exit configurations for the aircraft – which will have a high-density layout compared with the regular MAX 8 – also specifically referred to the variant as the 737-8200 when they were published for comment in 2018. 'Ryanair intends to take its version of the aircraft, which was also previously known as the 'MAX 200', with a 197-seat layout'.

More Revelations emerge

The bad news kept on coming. On 7 December 2019 The FAA sought to impose a $3.9 million fine on Boeing for failing to prevent the installation of defective slat

Above and left: half a world away from the politics and executive decisions of Boeing, one of their products, in this case SE-REU, a 737-76N(W) operated by SAS is seen back-tracking on the runway at Skiathos before turning around on the frying pan at the harbour end. The soft evening light always makes for some artistic images. *(all Phil McCrackin Collection)*.

Below: the view from the other direction as EI-EOJ, a 737-8BK(W) of air italy group swoops in to land at Skiathos.

tracks on about 130 737 NGs. The slat tracks were weakened by a condition known as hydrogen embrittlement that occurred during cadmium-titanium plating. A defective slat track could cause the slat to detatch and potentially strike the aircraft, resulting in injury to occupants and/or preventing continued safe flight and landing.

The FAA alleged that Boeing '...failed to adequately oversee its suppliers to ensure they complied with the company's quality assurance system. Boeing knowingly submitted aircraft for final FAA airworthiness certification after determining that the parts could not be used due to a failed strength test.'

Boeing said on 6 December 2019 that it had not been informed of any in-service issues related to the batch of slat tracks in question. It added that 'all affected 737 NGs have been inspected AD 2019-14-12 and all slat track installations determined to be required have been completed on the NGs.'

Boeing had thirty days to respond either by paying the fine or challenging it. It was thought by many industry insiders that the issue could also affect the MAX.

On Monday 16 December 2019 Boeing announced a temporary halt to 737 MAX production starting in January 2020 as it waited for the embattled airliner to be recertified to return to service.

The company's stock price closed down more than four per cent after early reports began to circulate that Boeing might halt the airliner's production, and then fell over another half per cent in after-hours trading following the company's official announcement.

Putting the troubled airliner back in the air was proving to be immensely difficult, causing major financial and reputational issues for Boeing.

Orders for the 737 MAX dried up, and it was not until November 2019 that Boeing recorded its first new orders since the grounding. In the meantime, the company continued to produce MAXs as it hoped for quick recertification by airline regulators around the globe.

That process, which faced a number of setbacks, had been pushed into 2020 and as there was an inventory of about 400 of the airliners in storage, the company said the continued uncertainty of the 737 MAX's future forced it to make the drastic move to pause the machine's production and shift its focus to delivering machines it had already produced. 'We believe this decision is least disruptive to maintaining long-term production system and supply chain health' Boeing said in a media release.

There was no indication about how long it expected production to be halted. Certainly, Boeing had been still hoping to get certification for the airliner to fly again before the end of 2019. However, FAA administrator Stephen Dickson said there was no chance that certification process could be completed before the end of the year, since the FAA - the US regulatory body - said that it would exercise full control over the certification of every 737 MAX aircraft built by Boeing. That meant that every MAX coming off the assembly line would need FAA's certification before being delivered to customers.

Boeing had been building forty-two of the 737 MAX airliners a month since the grounding, claiming it was so as not to cause hardship for its suppliers or be forced to lay off workers it would need later. Boeing said it did not expect to have to lay off or furlough employees. Affected workers would either continue 737-related work or be temporarily assigned to other teams.

The halt in deliveries was a huge cash drain to Boeing, which makes most of its revenue once an airliner is delivered. At the same time, though, the halt would reduce the company's expenses while deliveries were suspended. However, Boeing would

Russian state airline Aeroflot operated a number of Bahamian registered 737s, as represented here by VP-BON, a 737-8LJ(W) leased from Avia Capital Leasing (Rostechnology) and seen towards the end of its landing roll at Alicante Airport, Spain. *(Phil McCrackin Collection)*.

later have to pay billions of dollars to its airline customers to compensate them for the cost of the grounding.

'Safely returning the 737 MAX to service is our top priority,' a company said: 'We know that the process of approving the 737 MAX's return to service, and of determining appropriate training requirements, must be extraordinarily thorough and robust, to ensure that our regulators, customers, and the flying public have confidence in the 737 MAX updates. We remain fully committed to supporting this process.'

The company said it would provide financial information regarding the production suspension in January 2020, when it would announce its earnings for the last three months of 2019.

Photographs started to emerge of completed aircraft parked and components stacked outside assembly plants. Almost one hundred completed fuselages were sitting unused in the Wichita, Kansas, parking lot of Spirit AeroSystems, one of the hundreds of companies in the Boeing supply chain that help produce the finished product. Spirit AeroSystems had the capacity to produce about fifty fuselages a month for the MAX, and it employed about 13,000 people in Wichita. Fuselages were photographed on railcars parked in railway sidings at Wichita awaiting transport to Seattle. Other images showed completed airliners spilling over into the Boeing employee parking lot in Renton, Washington. The Southern California Logistics Airport - more commonly known as Victorville, saw many MAXs flown in and stored there, just as they were at Moses Lake in Washington State.

As of 1 January 2020, the exact number of undelivered MAX's had not been revealed, but was commonly thought to be between 380 and 420.

Boeing was frantically awaiting re-certification for the MAX by the FAA so it can return the airliner to the skies, but the FAA refused to commit to a time line for the return, instead repeatedly saying the plane would return when ready.

The FAA acknowledged that its staff was facing pressure to approve the airliner quickly but went on the record saying that it would ignore all factors except safety in deciding when the aircraft would be certified to carry passengers again.

The fall-out continues.

17 December 2019 - bases at Nuremberg and Stockholm Skavsta were to be shut down by Ryanair until at least the summer

of 2020 as a direct result of the 737 MAX crisis causing further delays in deliveries. By 4 February 2020, they were telling customers that they had pushed back its target to serve 200 million passengers a year until 2025 or 2026 following delays to the delivery of the group's first 737 MAX. The airline said it did not expect the aircraft to be delivered until September or October 2020, with cost savings likely not to be seen until late 2021.

In an update for investors, the airline said it been forced to revise its summer 2020 schedule again based on receiving just ten MAX aircraft rather than twenty as previously planned. CEO Eddie Wilson said Ryanair also expected to cut summer capacity in some other existing bases, and was in discussions with staff, unions and affected airports to finalise these 'minor reductions'.

American Airlines, a major customer for the MAX, again adjusted the expected service return date for the aircraft to April 2020 at the earliest and said it remained in continuous contact with the FAA, the Department of Transportation and Boeing.

With American's flight attendants said to be begging not to be rostered on to the MAX, IATA senior vice-president Gilberto López Meyer said a week previously that public and employee confidence in the aircraft was critical to a successful return to service.

'Airline accidents are extremely rare and certainly two accidents involving the same aircraft type within a few months of each other are even more of an extraordinary outcome' he added. Clearly he was talking things up on behalf of Boeing.

Clearly the troubled Boeing 737 MAX was continuing to pose safety concerns even after it is recertified and returned to service, according to a number of risk management specialists. They said it 'will almost certainly be safer than it was prior to the grounding. The safeguards that were designed to detect and eliminate design flaws failed in the design and certification of the plane. Neither Boeing nor the US FAA have outlined plans to re-examine the aircraft for other potential design and/or safety flaws.'

To nobody's real surprise, two days before Christmas 2019 Boeing announced that chief executive Dennis Muilenburg had

Bringing attendees to Zurich for the World Economic Forum at Davos in 2019 was N301SR, a 737-7VJ(BBJ) operated by Essar Shipping Ports & Logistics Limited, an Indian shipping corporation for the global energy business. *(Emma Dayle collection)*

Switzerland is usually a happy-hunting ground for Boeing Business Jets. HereVP-CEC, a 737-9HWER(W) starts its take-off roll from Geneva. Sources say it was being operated by Peridot Associated SA. *(Emma Dayle collection)*

stepped down 'with immediate effect' as it continued to deal with events swirling around the 737 MAX.

The manufacturer's board said, 'a change in leadership was necessary to restore confidence in the company moving forward as it works to repair relationships with regulators, customers and all other stakeholders'.

Muilenburg was in line to receive $26.5m in cash and stock as part of his exit package. His payout could reach as high as $58.5m, depending on how it would be structured, according to a US Securities and Exchange Commission filing, including a pension of $807,000 annually and Boeing stock worth another $13.3m.

It seems that Boeing chose not to fire the executive, allowing him to keep his pay package and so saving the company from the impression that its most senior executive had done something wrong.

CFO Greg Smith would act as interim chief until current chairman David Calhoun took up the role on 13 January.

The change in leadership came as US lawmakers continued to question Boeing and the FAA about what happened during the initial certification process for the 737 MAX, which was still grounded. It also opened the door to the possibility of resetting its relationship with the US regulator. The FAA's administrator, Steve Dickson, had made clear his dissatisfaction with some of Boeing's actions during the MAX grounding, and he called a face-to-face meeting with Muilenburg and new Boeing Commercial Airplanes President Stan Deal on 12 December to express his concerns. During the meeting, Dickson called out the company's unrealistic return-to-service timeline that it maintained throughout the second half of 2019, even as the FAA insisted it has no defined time frame for its review. Dickson, sworn in as the top FAA official on 11 August, several weeks after Boeing began projecting a late 2019 return-to-service, also told Boeing the quality and timeliness of its MAX-related paperwork was lacking, according to a summary of the meeting sent to US lawmakers.

Boeing was desperately trying to save its image in the face of the tragedies, apologising at every opportunity for the deaths and reiterating that it was making changes to its internal processes to ensure a similar incident never happens again.

Commenting on Boeing's decision to oust Muilenburg, board member and new non-executive chairman Lawrence Kellner said: 'On behalf of the board of directors, I am pleased that Dave has agreed to lead Boeing at this critical juncture. Dave has deep industry experience and a proven track record of strong leadership, and he recognises the challenges we must confront. The board and I look forward to working with him and the rest of the Boeing team to ensure that today marks a new way forward for our company.'

Calhoun added: 'I strongly believe in the future of Boeing and the 737 MAX. I am honoured to lead this great company and the 150,000 dedicated employees who are working hard to create the future of aviation.'

Whether more management heads would roll at Boeing remained to be seen now that Dennis Muilenburg had been fired as chief executive, but the era of Boeing striving to be a global industrial champion and putting shareholders over other priorities probably had come to an end.

Meanwhile, Larry Kellner, a board member and former CEO of Continental Airlines, became the new Boeing non-executive chairman, effective immediately. Calhoun had served as non-executive chairman since October, when the Boeing board suddenly stripped Muilenburg of the chairman title. Among other things, the latest announcement confirms a separation of chairman and CEO, as corporate governance advocates and some Boeing critics have called for all year.

'Under the company's new leadership, Boeing will operate with a renewed commitment to full transparency, including effective and proactive communication with the FAA, other global regulators and its customers,' the prepared statement continued.

Analysts note that even after the airliner has been given the regulatory green light, it could take a year until the grounded fleet are all back in the air. Boeing's manufacturing shutdown looks sure to inflate the $9bn (£7bn) it has already cost the planemaker, including $5bn set aside to compensate airline customers for the grounding. Ronald Epstein, analyst at Bank of America Merrill Lynch, estimated the 737 MAX grounding will eventually cost Boeing about $14bn.

Airlines have withdrawn it from their schedules for several months. American Airlines pulled it until at least 7 April 2020 and Southwest, the biggest user of the airliner, said the flying ban had cut its operating income by $830m this year.

Other customers have launched legal action against Boeing. Timaero, an Irish company that sells and leases aircraft, has filed a lawsuit in the US seeking to cancel

Bahamian registered 737-9LBER(W) VP-BDB seen about to depart Geneva on a very cold afternoon in 2017. The aircraft's 'registered keepers' were the Dallah Albaraka Group. *(Emma Dayle collection)*

The IAI 737-700 BDSF - Israel Aerospace Industries developed a supplemental type certificate for a 737-700BDSF cargo conversion. A first prototype aircraft is seen here just after conversion from passenger to full freighter configurations at IAI's facility. The conversion includes the installation of a cargo door on the fuselage, additional structural modification to support full load capacity, and implementation of smoke and fire detection in the cargo bay. *(Dr Harry Friedman Collection)*

an order for twenty-two MAX machines and damages of at least $185m. Documents filed in a US district court alleged that Boeing had acted fraudulently in selling the troubled aircraft. Turkish Airlines, which has twenty-four of the aircraft, is also believed to be preparing legal action.

2020 Vision...

2020 may stand for perfect vision, but just six days into the year two things came to light that the company may not have wanted to see. During the exhaustive review of the 737 MAX systems demanded by safety regulators, it became clear that Boeing had discovered a potential new wiring-design problem that almost certainly would further delay the company's target of returning the airliner into service.

At the time of discovery, the impact of the potential design problem inside some wiring bundles was not known.

According to insiders, Boeing engineers discovered what was termed 'a theoretical possibility' of an electrical short circuit in wires connected to the airliner's moveable horizontal tail that could cause the tail to swivel uncommanded by the pilot, pushing the nose down.

Although this potential fault in the wiring bundles was unrelated to the flight-control system on the two crash flights of the MAX, Boeing said that the discovery was part of 'a robust and thorough certification process to ensure a safe and compliant design', and that it was analyzing the risk that it could produce a similar outcome to the crash scenarios.

'We identified this wire bundles issue as part of that rigorous process, and we are working with the FAA to perform the appropriate analysis,' said Boeing spokesman Gordon Johndroe. 'It would be premature to speculate as to whether this analysis will lead to any design changes'.

One Boeing insider said it was not yet clear if this is a real concern. 'Our current understanding is that analysis may show that the theoretical fault cannot occur in the specific way required, and that other protections already in place – ranging from shielding to insulation to circuit breakers – would prevent it from being possible'.

During the review, Boeing's engineering team identified a theoretical scenario in which three wires routed close together might cause an electrical short that could result in a high-speed, continuous,

TonleSap Airlines' XU-TSA, a 737-33A. *(Emma Dayle collection)*

horizontal stabiliser runaway.

During the original certification of the MAX, Boeing's system safety assessment classified such a runaway stabiliser - in which the horizontal tail swivels without pilot input to push down the nose of the jet - as a major hazard, meaning a flight upset that would likely cause only minor injuries to those on board.

Given the outcome of the two crashes, when both flight crews failed to cope with a runaway stabiliser, the event was reassessed last summer as 'catastrophic' — two hazard levels higher, signifying a risk of losing the plane. A catastrophic hazard classification requires a design such that no single failure could trigger the event. That led to re-examining every possible system failure and the discovery of the potential electrical short.

This fault was not related to the MCAS flight control software - if further analysis confirmed that the wiring was a real concern, the fix would require additional separation of a short segment of two wire bundles, the Boeing insider said.

Adding more wire separation toward the tail of the airliner is difficult because the structure narrows there and there is little extra room for manoeuvre.

Even as Boeing engineers tried to figure out if the failure could really happen, they were also working on designing the separation.

A separate manufacturing issue affecting MAXs built in 2019 will require hours of repair work on a large number of jets to

UTair's VP-BXO, a 737-524(W) leased from Continental Airlines. *(Hank Warton Collection)*

ensure the engines are fully protected from a lightning strike before they can fly again. Manufacturing flaws left MAXs vulnerable to lightning.

On any jet aircraft, the two areas that draw the most lightning strikes are the nose of the aircraft and the pods surrounding the engines that thrust out ahead of the wing. On the 737, the engine pods and the pylons that hold them to the wing, which are largely made of nonconducting carbon fibre composite, have a metal foil just beneath the surface to safely disperse the current from such a strike.

Boeing said in a statement that an incorrect manufacturing procedure used on some MAXs damaged the protective metal foil on two panels covering the engine pylons. 'On some airplanes built between February 2018 and June 2019, the protective foil inside the composite panels may have gaps,' Boeing said.

Then came another bombshell reversal. In a release on Tuesday 7 January it was recommended that pilots receive training in a flight simulator before the grounded 737 MAX returns to flying. This was a complete reversal of the company's long-held position that computer-based training alone was adequate.

The recommendation was based on changes to the airliner, test results and a commitment to the safe return of the MAX. Boeing's interim chief executive, Greg Smith, said in a statement that Boeing decided to recommend simulator training because of the importance to Boeing of gaining public and airline confidence in the MAX. The final decision on the nature of training would be up to regulators, including the FAA, who stated that it in turn would consider Boeing's recommendation but also would rely on upcoming tests using pilots from US and foreign airlines.

Those tests were designed to help regulators determine flight training and emergency procedures, said FAA spokesman Lynn Lunsford. 'The FAA is following a thorough process, not a set timeline, ensuring that any design modifications to the 737 MAX are integrated with appropriate training and procedures.'

Boeing had long held that pilots who could fly older 737s only needed a computer course, roughly an hour-long, on a tablet to fly the MAX. That helped airlines avoid timely and costly training in simulators.

Boeing even offered to pay Southwest Airlines a rebate of $1m per aircraft if the airline had to train its nearly 10,000 pilots in simulators before they could fly the MAX.

THY Turkish Airlines TC-JYJ, a 737-9F2ER(W). *(Emma Dayle collection)*

Boeing's statement was countered by Peter DeFazio the same day: 'While I agree with Boeing's decision to recommend that pilots undergo simulator training on the MAX, it's remarkable that it took two deadly crashes, numerous investigations and untold public pressure before Boeing arrived at this decision.

'While our Committee continues to comb through records, conduct interviews, and hold public hearings to better understand how the system failed so drastically and ultimately led to the tragic deaths of 346 people, it's already abundantly clear that from its inception, Boeing's business model for the 737 MAX was premised on Boeing's unreasonable, cost-saving assurance to airlines that pilots qualified to fly a different 737 variant, the 737 Next Generation, should not undergo simulator training to fly the 737 MAX. Boeing made a fundamentally flawed decision that put production and profits ahead of the public's safety. Safety must always come first in the aviation industry. It will be incumbent upon Boeing to ensure that the recommended simulator training is sufficient to provide all flight crews around the globe with both the proficiency and the information they need to fly the 737 MAX safely.'

Rick Larsen also provided an insight: 'The 737 MAX should not return to service until the FAA determines it is safe to do so. As Chair of the Aviation Subcommittee, I remain committed to the thorough oversight of the 737 MAX certification process and ensuring the aircraft's safe return to service. I remain concerned that Boeing initially had financial incentives to avoid simulator training and believe today's recommendation that all 737 MAX pilots receive simulator training is long overdue. I will continue to keep the 346 victims of the two tragic 737 MAX crashes and their families at the forefront of the Committee's investigation, as well as the dedicated women and men of Boeing who design, assemble and build the aircraft.'

'Designed by clowns... Supervised by monkeys'.

Later that week the cascade of revelations turned into an avalanche. Reports surfaced in the media that Boeing had expressed regret at a raft of embarrassing communications it sent to investigators the previous day: 'These communications contain provocative language, and, in certain instances, raise questions about Boeing's interactions with the FAA in connection with the simulator qualification process,' the company said in a statement to Congress. 'Having carefully reviewed the issue, we are confident that all of Boeing's MAX simulators are functioning effectively.'

'We regret the content of these communications, and apologize to the FAA, Congress, our airline customers and to the flying public for them. The language used in these communications, and some of the sentiments they express, are inconsistent with Boeing values, and the company is taking appropriate action in response. This will ultimately include disciplinary or other personnel action, once the necessary reviews are completed.'

So what were the comments? One, dating from 2017 in which one employee said that 'this airplane is designed by clowns, who are in turn supervised by monkeys.' Of course, the media loved it. Boeing employees mocked federal rules, talked about deceiving regulators and joked about potential flaws in the 737 MAX as it was being developed, according to over a hundred pages of internal messages delivered to congressional investigators.

'I still haven't been forgiven by God for the covering up I did last year,' one of the employees said in messages dating from 2018, apparently in reference to interactions with the FAA.

'Would you put your family on a MAX simulator trained aircraft? I wouldn't,' said another employee to a colleague in another exchange from 2018, before the first crash.

The employees appear to have discussed instances where the company concealed such problems from the FAA during the regulator's certification of the simulators, which were used in the development of the MAX, as well as in training for pilots who had not previously flown a 737.

The release of the communications - both emails and instant messages - was the latest embarrassing episode for Boeing in a crisis that has cost the company billions of dollars and wreaked havoc on the aviation industry across the globe.

The messages threatened to further complicate Boeing's tense relationship with the FAA. Both the company and agency indicated that the messages raised no new safety concerns, but they echoed troubling internal communications among Boeing employees that were previously made public.

In several instances, Boeing employees insulted the FAA officials reviewing the airliner. Numerous employees seemed consumed with limiting training for airline crews to fly the airliner, a significant victory for Boeing which would benefit the company financially.

In an email from August 2016, a marketing employee at the company cheered the news that regulators had approved a short computer-based training for pilots who have flown the 737 NG, the predecessor to the MAX, instead of requiring simulator training. 'You can be away from an NG for thirty years and still be able to jump into a MAX? LOVE IT!!' the employee says, following up later with an email noting: 'This is a big part of the operating cost structure in our marketing decks.'

For those uninitiated in the delights of public relation-speak, 'Marketing decks' are visual presentations used by marketers, public relations managers and advertising

Okay Airways of China took delivery of the 9999th 737, the significance of which was explained by Erik Nelson, vice president of Boeing 737 Field Operations and Delivery, who said that 'the figure of 9999 means eternity in Chinese culture, and that Boeing and the Chinese air industry will continue to deepen their long-term relations of friendship and cooperation'. *(Wan Hunglo Collection)*

4X-EHC, a 737-985ER(W) of El Al taxies in at Zurich. *(Emma Dayle collection)*

executives for one of two purposes:either as a tool for selling a product or service to a client, or as a snapshot of a given time period in the company's marketing and advertising programme. The messages outraged several lawmakers, who saw a total disregard for safety and broader problems with the culture at the company.

Senator Richard Blumenthal, [D] of Connecticut, said in an interview that he would push for new congressional hearings to question Boeing leadership about the 'astonishing and appalling' messages.

Boeing explained that it notified the FAA about the documents in December and that it had 'not found any instances of misrepresentations to the FAA with its simulator qualification activities,' despite the employee's comment about 'covering up' issues with the simulator.

Lynn Lunsford, a spokesman for the FAA, said in a statement that the messages did not reveal any new safety risks. 'Upon reviewing the records for the specific simulator mentioned in the documents, the agency determined that piece of equipment has been evaluated and qualified three times in the last six months. Any potential safety deficiencies identified in the documents have been addressed. While the tone and content of some of the language contained in the documents is disappointing, the FAA remains focused on following a thorough process for returning the Boeing 737 Max to passenger service.'

House Committee on Transportation and Infrastructure Chair Peter DeFazio, who was investigating the 737 MAX affair, issued a statement on 9 January 2020: 'These newly-released emails are incredibly damning. They paint a deeply disturbing picture of the lengths Boeing was apparently willing to go to in order to evade scrutiny from regulators, flight crews, and the flying public, even as its own employees were sounding alarms internally. I can only imagine how painful it must be for the families of the 346 victims to read these new documents that detail some of the earliest and most fundamental errors in the decisions that went into the fatally flawed aircraft'.

'For nearly ten months, my Committee has been investigating the design, development and certification of the MAX, and in that time, our investigation has uncovered multiple, serious problems with Boeing's decision-making and the priority that was placed on production and profit over safety. But these new emails bring my concerns to an entirely new level. They show a coordinated effort dating back to the earliest days of the 737 MAX program to conceal critical information from regulators and the public. While it is also clear from these emails that the problems did not merely stem from a lone Boeing employee who uses colorful language in his communications, I have reiterated my request for an interview with former Boeing 737 MAX Chief Technical Pilot Mark Forkner to his attorneys, and I expect to hear from them at the earliest possible date.'

A change of tone?

21 January, Chicago and a press release from Boeing Communications that really was of a very different tone. 'As we have emphasized, the FAA and other global regulators will determine when the 737 MAX returns to service. However, in order to help our customers and suppliers plan their operations, we periodically provide them with our best estimate of when regulators will begin to authorize the ungrounding of the 737 MAX'.

'We are informing our customers and suppliers that we are currently estimating that the ungrounding of the 737 MAX will begin during mid-2020. This updated estimate is informed by our experience to date with the certification process. It is subject to our ongoing attempts to address known schedule risks and further developments that may arise in connection with the certification process. It also accounts for the rigorous scrutiny that regulatory authorities are rightly applying at every step of their review of the 737 MAX's flight control system and the Joint Operations Evaluation Board process which determines pilot training requirements'.

'Returning the MAX safely to service is our number one priority, and we are confident that will happen. We acknowledge and regret the continued difficulties that the grounding of the 737 MAX has presented to our customers, our regulators, our suppliers, and the flying public. We will provide additional information about our efforts to safely return the 737 MAX to service in connection with our quarterly financial disclosures next week'.

This told the world little more than they already knew, apart from announcing a further delay in the return to service. What was different was the second part of the release, warning of 'Caution Concerning Forward-Looking Statements'.

'Certain statements in this release may be 'forward-looking' within the meaning of the Private Securities Litigation Reform Act of 1995. Words such as 'may,' 'should,' 'expects,' 'intends,' 'projects,' 'plans,' 'believes,' 'estimates,' 'targets,' 'anticipates,' and similar expressions generally identify these forward-looking statements. Examples of forward-looking statements include statements relating to our future financial condition and operating results, as well as any other statement that does not directly relate to any historical or current fact. Forward-looking statements are based on expectations and assumptions that we believe to be reasonable when made, but that may not prove to be accurate. These statements are not guarantees and are subject to risks, uncertainties, and changes in circumstances that are difficult to predict.

Many factors could cause actual results to differ materially and adversely from these forward-looking statements, including the timing and conditions surrounding the return to service of the 737 MAX fleet and the resumption of 737 MAX production, general conditions in the economy and our industry, including those due to regulatory changes, our reliance on our commercial airline customers and our suppliers, and changes in our accounting estimates, as well as the other important factors disclosed previously and from time to time in our filings with the Securities and Exchange Commission. Any forward-looking

From the snowy Arctic to the warmth of the Caribbean, the 737 found use in every scenario.

Above: A Nolinor 737-200 lands on a snow-covered runway.

Below: 9Y-ANU of Caribbean Airlines back-tracks after landing at Queen Juliana Airport, St Maartin.

(both Hank Warton Collection)

statement speaks only as of the date on which it is made, and we assume no obligation to update or revise any forward-looking statement, whether as a result of new information, future events, or otherwise, except as required by law'.

This news all but ensured that northern hemisphere MAX operators would not have the aircraft before the peak mid-year travel period ended.

The move meant that initial revenue-service operations could not start until September or later - weeks longer than the most conservative airline estimates. US airlines would be among the first to have regulatory clearance to put MAXs back into service, but executives with US carriers estimated that they need between

USAF C-40B 10041 seen just at the point of touchdown. *(Hank Warton Collection)*

thirty and sixty days from the FAA's approval to prepare MAXs for service, train pilots and work the aircraft back into flight schedules.

Airlines in other countries would have to wait for their own regulators to lift the groundings, with some agencies pledging to take independent looks at both Boeing's work and FAA's approvals. It is not clear whether the ongoing FAA review gave other regulators time to complete some of their independent work, or if any timeline slippage in the US would be mirrored around the world.

The next major step on the MAX's approval path would have to be a certification flight, which was not scheduled. The latest projection was not driven by any new findings linked to Boeing's MAX changes and training revisions, but reflected the reality of the certification process's pace and accounts for some additional schedule slippage.

By the end of January 2020 the fallout from the production halt with the 737 MAX revealed that it was being dealt with by aerospace suppliers in two different ways. Some announced they were cutting numbers of employees, while others announced they were not, instead getting caught up with their backlog of work.

It was the time where many public companies began reporting 2019 financial results; top executives from United Technologies and Moog said they did not intend to lay off workers over the MAX halt. UTC was assuming a ninety-day halt based on guidance from Boeing, UTC Chairman and CEO Greg Hayes told the media on 28 January. 'We do not anticipate any layoff. I think that would be the easiest thing to do, but quite frankly, given the scarcity of talented aerospace workers out there, we're not going to be laying anybody off for a ninety-day delay here. I think we're going to work on the backlog.'

But the day before, Arconic CEO John Plant told analysts he expected to make headcount reductions, as well as to explore partially paid vacations and worker shift changes as his company tried to mitigate the MAX halt. Earlier in the month, Spirit AeroSystems announced at least 2,800 layoffs due to the MAX.

Compounding the MAX effect on the supply chain was that several large providers were in the throes of major mergers or divestitures - corporate changes that often lead to workforce reductions by themselves. Arconic were scheduled to split into two in early April 2020, becoming Howmet Aerospace and rolled-aluminium provider Arconic.

End of year figures.

Then, on 29 January 2020 Boeing reported its first annual loss since 1997 as the cost of the 737 MAX crisis continued to climb. The company reported a net loss of $636 million for 2019, compared to a $10.5 billion profit it made in 2018.

Sooner than take extracts of the reports from other sources, I have decided to quote the entire section verbatim - 'straight from the horse's mouth' as it were - and use tables, charts and detailed extracts from Boeing's own fourth-quarter results.

Boeing Reports Fourth Quarter Results

Fourth Quarter 2019

- Financial results continue to be significantly impacted by the 737 MAX grounding
- Revenue of $17.9 billion, GAAP loss per share of ($1.79) and core (non-GAAP)* loss per share of ($2.33)

Full-Year 2019

- Revenue of $76.6 billion, GAAP loss per share of ($1.12) and core (non-GAAP)* loss per share of ($3.47)
- Operating cash flow of ($2.4) billion; cash and marketable securities of $10.0 billion
- Total backlog of $463 billion, including over 5,400 commercial airplanes

Table 1. Summary Financial Results

(dollars in Millions, except per share data)

	Fourth Quarter			Full Year		
Revenues	2019	2018	Change	2019	2018	Change
	$17,911	$28,341	(37)%	$76,559	$101,127	(24)%
GAAP*						
(Loss)/Earnings From Operations	($2,204)	$4,175	NM	($1,975)	$11,987	NM
Operating Margin	(12.3)%	14.7%	NM	(2.6)%	11.9%	NM
Net (Loss)/Earnings	($1,010)	$3,424	NM	($636)	$10,460	NM
(Loss)/Earnings Per Share	($1.79)	$5.93	NM	($1.12)	$17.85	NM
Operating Cash Flow	($2,220)	$2,947	NM	($2,446)	$15,322	NM
Non-GAAP**						
Core Operating (Loss)/Earnings	($2,526)	$3,867	NM	($3,390)	$10,660	NM
Core Operating Margin	(14.1)%	13.6%	NM	(4.4)%	10.5%	NM
Core (Loss)/Earnings Per Share	($2.33)	$5.48	NM	($3.47)	$16.01	NM

* = Generally Accepted Accounting Principles. ** = Boeing's own non-GAAP measures.

NM = Not Meaningful.

The Boeing Company reported fourth-quarter revenue of $17.9 billion, GAAP loss per share of ($1.79) and core loss per share (non-GAAP) of ($2.33), primarily reflecting the impacts of the 737 MAX grounding as shown in Table 1. Boeing recorded operating cash flow of ($2.2) billion and paid $1.2 billion of dividends.

'We recognize we have a lot of work to do,' said Boeing President and Chief Executive Officer David Calhoun. 'We are focused on returning the 737 MAX to service safely and restoring the long-standing trust that the Boeing brand represents with the flying public. We are committed to transparency and excellence in everything we do. Safety will underwrite every decision, every action and every step we take as we move forward. Fortunately, the strength of our overall Boeing portfolio of businesses provides the financial liquidity to follow a thorough and disciplined recovery process.'

Table 2. Cash Flow

(Millions)	Fourth Quarter 2019	Fourth Quarter 2018	Full Year 2019	Full Year 2018
Operating Cash Flow	($2,220)	$2,947	($2,446)	$15,322
Less Additions to Property, Plant & Equipment	($447)	($495)	($1,834)	($1,722)
Free Cash Flow	($2,667)	$2,452	($4,280)	$13,600

Operating cash flow was ($2.2) billion in the quarter, primarily reflecting the impact of the 737 MAX grounding as well as timing of receipts and expenditures - see Table 2. During the quarter, the company paid $1.2 billion of dividends.

Table 3. Cash, Marketable Securities and Debt Balances

(Billions)	Quarter-End Q4 19	Quarter-End Q3 19
Cash	$9.5	$9.8
Marketable Securities	$0.5	$1.1
Total	$10.0	$10.9
Debt Balances:		
The Boeing Company, net of intercompany loans to BCC	$25.3	$22.8
Boeing Capital, including intercompany loans	$2.0	$1.9
Total Consolidated Debt	$27.3	$24.7

Marketable securities consists primarily of time deposits due within one year classified as 'short-term investments.'

Cash and investments in marketable securities totaled $10.0 billion, compared to $10.9 billion at the beginning of the quarter as shown in Table 3. Debt was $27.3 billion, up from $24.7 billion at the beginning of the quarter primarily due to increased commercial paper borrowings.

Total company backlog at quarter-end was $463 billion and included net orders for the quarter of $13 billion.

Table 4. Commercial Airplanes
(dollars in Millions)

	Fourth Quarter 2019	Fourth Quarter 2018	Fourth Quarter Change	Full Year 2019	Full Year 2018	Full Year Change
Commercial Airplanes Deliveries	79	238	(67)%	380	806	(53)%
Revenues	$7,462	$16,531	(55)%	$32,255	$57,499	(44)%
(Loss)/Earnings from Operations	($2,844)	$2,600	NM	($6,657)	$7,830	NM
Operating Margin	(38.1)%	15.7%	NM	(20.6)%	13.6%	NM

Commercial Airplanes fourth-quarter revenue was $7.5 billion and fourth-quarter operating margin decreased to (38.1) percent reflecting lower 737 deliveries and an additional pre-tax charge of $2.6 billion related to estimated potential concessions and other considerations to customers related to the 737 MAX grounding (Table 4). The estimated costs to produce 737 aircraft included in the accounting quantity increased by $2.6 billion during the quarter, primarily to reflect updated production and delivery assumptions. In addition, the suspension of 737 MAX production and a gradual resumption of production at low production rates will result in approximately $4 billion of abnormal production costs that will be expensed as incurred, primarily in 2020.

Commercial Airplanes delivered 79 airplanes during the quarter, including 45 787's, and captured orders for 30 737 MAX aircraft at the Dubai Air Show and two 777 freighters for Lufthansa. The 787 program also booked 36 net orders in the quarter. As previously announced, the 787 production rate will be reduced from the current rate of fourteen airplanes per month to twelve airplanes per month in late 2020. Based on the current environment and near-term market outlook, the production rate is expected to be further adjusted to ten airplanes per month in early 2021, and return to twelve airplanes per month in 2023. The first flight of the 777X was completed on January 25, and first delivery is targeted for 2021.

Commercial Airplanes backlog included over 5,400 airplanes valued at $377 billion.

Table 5. Defense, Space & Security

(dollars in Millions)

	Fourth Quarter			Full Year		
	2019	2018	Change	2019	2018	Change
Revenues	$5,962	$6,874	(13)%	$26,227	$26,392	(1%)
Earnings from Operations	$31	$771	(96)%	$2,608	$1,657	57%
Operating Margin	0.5%.	11.2%	(10.7) Pts.	9.9%	6.3%.	3.6 Pts

Defense, Space & Security fourth-quarter revenue decreased to $6.0 billion primarily driven by lower volume across the portfolio as well as the impact of a Commercial Crew charge shown in Table 5. Fourth-quarter operating margin decreased to 0.5 per cent due to a $410 million pre-tax Commercial Crew charge primarily to provision for an additional uncrewed mission for the Commercial Crew program, performance and mix. NASA is evaluating the data received during the December 2019 mission to determine if another uncrewed mission is required.

During the quarter, Defense, Space & Security received an award for ten Space Launch System core stages and up to eight Exploration Upper Stages. Defense, Space & Security also received contracts for the remanufacture of 47 AH-64E Apache helicopters for three countries and to upgrade the NATO Airborne Warning & Control System fleet. Significant milestones achieved during the quarter included the delivery of the first modified MV-22 Osprey to the U.S. Marine Corps and delivery of the first P-8A Poseidon aircraft to the United Kingdom Royal Air Force. Defense, Space & Security also conducted a Commercial Crew spacecraft uncrewed Orbital Flight Test.

Backlog at Defense, Space & Security was $64 billion, of which 29 percent represents orders from customers outside the U.S.

Table 6. Global Services

(dollars in Millions)

	Fourth Quarter			Full Year		
	2019	2018	Change	2019	2018	Change
Revenues	$4,648	$4,908	(5)%	$18,468	$17,056	8%
Earnings from Operations	$684	$737	(7)%	$2,697	$2,536	6%
Operating Margin	14.7%	15.0%.	(0.3) Pts.	14.6%.	14.9%.	(0.3) Pts

Global Services fourth-quarter revenue was $4.6 billion, primarily driven by lower commercial services volume (Table 6). Fourth-quarter operating margin decreased to 14.7 percent primarily due to a charge related to the retirement of the Aviall brand and mix of products and services, partially offset by a gain on divestiture.

During the quarter, Global Services was awarded V-22 support contracts for Japan and the U.S. and AH-64 and CH-47 global support for the U.S. Army. Global Services signed a multi-year Landing Gear Exchange services agreement with LATAM Airlines Group and a 5-year digital navigation renewal agreement with Saudi Arabian Airlines. Global Services also expanded its digital offerings by launching ForeFlight Dispatch and signed a contract with Flexjet to be the inaugural customer.

Table 7: The Boeing Company and Subsidiaries Operating and Financial Data (Unaudited)

	Twelve months ended		Three months ended	
Deliveries	December 31		December 31	
Commercial Airplanes	2019	2018	2019	2018
737	127	58	9	173
747	7	6	2	1
767	43	27	11	14
777	45 (2)	48	12 (1)	11
787	158	145	45	39
Total	380	806	79	238
Note: Aircraft accounted as revenues by BCA and as operating leases in consolidation identified by parentheses				
Defense, Space & Security				
AH-64 Apache (New)	37	—	10	—
AH-64 Apache (Remanufactured)	74	23	18	11
C-17 Globemaster III	1	—	—	—
C-40A	2	—	—	—
CH-47 Chinook (New)	13	13	—	2
CH-47 Chinook (Renewed)	22	17	6	3
F-15 Models	11	10	4	2
F/A-18 Models	23	17	7	7
KC-46 Tanker	28	—	7	—
P-8 Models	18	16	4	6
Commercial and Civil Satellites	2	1	1	—
Military Satellites	—	1	—	1

Table 8: Total backlog (Dollars in millions)

	December 31 2019	December 31 2018
Commercial Airplanes	$376,593	$408,140
Defense, Space & Security	63,908	61,277
Global Services	22,902	21,064
Total backlog	$463,403	$490,481
Contractual backlog	$436,473	$462,070
Unobligated backlog	26,930	28,411
Total backlog	$463,403	$490,481

Chapter Fourteen

Can We Fix It?

There has been much talk of the problem with the MAX being MCAS and its associated hardware. That may well be the current problem, but in engineering terms of cause and effect, it is not the primary one. The so-called 'fixing' of the MCAS software is only the latest in a long line of applying 'patches' in both the IT and physical sense to work around the original, fundamental problem. The 'linage of modification' when working backwards goes like this.

Question: MCAS is faulty - how is that fixed?

Answer: Use updated software and additional sensors, so it performs properly.

Question: But why is MCAS fitted in the first place?

Answer: the LEAP B engines installed and their installation plus aerodynamic nacelles cause handling abnormalities at certain speeds.

Question: Then why fit the LEAP engines?

Answer: To obtain better efficiency in order to get sales.

Question: Could they have been installed differently?

Answer: No, there were ground clearance issues created by the length of the landing gear legs.

Question: Could the landing gear legs be moved outwards and then extended to improve ground clearance?

Answer: True, the track of the landing gear was increased from seventeen feet two inches with the series 200 to eighteen feet nine with the series 800, a potential increase of nine inches a side, but any further imrpovement could not take place without significant modification to the aircraft fuselage, wings, engine supports and flaps.

Question: So why not undertake those changes?

Answer: To undertake such significant modifications would be beyond what could be called alterations to an existing design. It is in effect, creating a different airliner, and that would require a complete testing and certification process from a blank sheet of paper, an extremely time consuming and expensive process that would make whatever was designed significantly disadvantaged to rival machines such as the A320 NEO.

So it is therefore that the origins of the so-called 'problems' with the design of the Boeing 737 MAX can be traced back to before metal was cut on the original 737-100 when the proximity of the Boeing 737 to the ground enabled the aircraft and its successors to shorten their turnaround times. The airliner was designed from the onset with 'short legs' to allow it to be efficiently used at second-ranked airports with limited stair systems. Ground clearance was never a problem with the P&W JT-8 'stove-pipe' engines.

However, with the advent of the CFM-56s, the larger diameter high-bypass engines caused problems with ground clearance, requiring some of the accessories to be moved around to the sides of the engine, to allow a flat bottom to be created, given the nacelle something of a 'hamster-pouch' look - technically called an 'ovoid' shape - when viewed from the front.

This may sound like stating the obvious, but ground clearance is determined by the length of the landing gear legs, and on the 737 these were positioned so that for the given length required they would fit into open-faced 'wells' in the aircraft centre-section. The geometry of this locked everything into a certain position.

To lengthen the landing gear would have required major re-design and strengthening of the main wing spar, for it is not possible to lengthen the legs and keep them in the same position, as they would hit each other when in the retracted state. Moving them outboard - towards the wingtips - would mean they would be closer to the engines, which had already grown in diameter, which in turn meant that they would then have to be moved and the wing structure re-designed to accommodate all the changes. It would be a very expensive and time-consuming process.

Boeing engineers were slowly but surely boxing themselves in, reducing options the

Below: This montage of an Airbus A320 on the left, and a Boeing 737 on the right clearly shows the difference in ground clearance and landing gear leg length. The montage also shows how the engines on the A320 have room to hang below the wing, whereas the 737 engines have to be positioned in front of the leading edge.

Opposite page: The three basic types of engine as fitted to the 737 that had ever-increasing diameters. Top, the Pratt & Witney JT-8D. Middle: the CFM-56, and Bottom: the CFM Leap 1B. Note also how over time the engines have 'moved up and forward' so as to maintain ground clearance.

available to them without starting on a clean sheet of paper. They had only one 'direction' left open to them - move the engines forward, which would, in turn, allow a small amount of upward movement. These changes were an acceptable way out with the -300 to -900s, but when it came to fitting the even larger diameter LEAP engines, the engineers looked at 'angling' them slightly. This unbalanced the thrust line, and as the engines were more powerful, this generated an undesirable nose up tendency at high thrust levels, especially during take off.

Ironically, the very same ingenious design became a challenge for the designers when the decision was made to launch the re-engined 737 MAX with bigger and more fuel-efficient CFM Leap-1B engine. As a result of the ground clearance constraint, Boeing and CFM opted for a customised core with either a sixty-six inch or sixty-eight-inch engine fan size. Engine fan size is important as it drives the propulsive efficiency and bypass ratio which have an impact on its specific fuel consumption (SFC). Every inch increase in the fan size generally leads to a half-a-percent reduction in the engine's fuel burn. However, a bigger engine also brings more drag and carries more weight, which negatively impacts on an engine's SFC. Boeing contended that the sixty-eight-inch engine fan size was the optimal choice, which balanced fuel burn saving, weight and drag of the engine.

Boeing Commercial Airplanes president and chief executive Jim Albaugh explained to delegates at a Goldman Sachs Global Industrials Conference in New York that the '... sixty-eight-inch fan size is really a sweet spot for us, a sweet spot in terms of fuel burn, drag on the airplane and also the additional weight added to the airplane. We did not just look at the engine, we looked at it as an integrated solution. Fan size is important, bypass ratio also drives weight and drives drag, I think we have made the right decision for us'.

In a teleconference with the media to back up the conference statements, Boeing 737 chief programme engineer John Hamilton explained that the weight of the airliner itself also drove the thrust requirement of the engine and played a role in the engine fan size decision as well. 'The 737-900ER is 10,000 pounds lighter than the A321. If you look at the operating weight per seat, our -900ER is nearly fifty pounds lighter per seat and so there is a much better structurally efficiency that goes into the 737 design than into the Airbus design' Hamilton asserted.

'Weight drives a lot of cost into an airline's operation. In addition to the fuel that it takes to lift that weight off the ground and carry it to the air, it also plays a part into maintenance cost and landing cost as well as the thrust requirement for the engine. And so Airbus on an A321 has to put 32,000 or 33,000 pounds of thrust on there versus -900ER it is only 26,000 pounds or 27,000 pounds, so a lot lower thrust requirement on the 737 programme'.

'Today our engine is seven inches smaller than the Airbus and yet we have a lower operating cost than the Airbus product. Again, this gives back to the structural efficiency of the airplane and the higher thrust requirement and the higher maintenance cost that the Airbus engine requires. As we size up the equivalent inch on the 737 MAX, Airbus is going to have to size up to seventy-eight inches on the A320 NEO's CFM Leap engine to provide the same sort of efficiency'.

'Both the fuel saving that comes from the engine as well as the drag associated with that engine as it flies through the air. And so you can think of a seventy-eight-inch engine is kind of like your Mack truck driving down the road and a sixty-eight inch engine as being a lot leaner and less drag on the engine and also the weight of the engine offsets the benefit as well. So when you look at drag, fuel efficiency, and the weight the sixty-eight-inch fan is really the right optimum solution for the 737 airplane going forward,' Hamilton emphasised.

Boeing said it had chosen a sixty-eight inch fan size for the aircraft's CFM International LEAP-1B engine, which, combined with improved aerodynamics through a revised 787-styled tail cone design, would deliver a ten-to-twelve per cent fuel burn saving over the existing 737 NG, as well as a four per cent lower fuel burn per seat and a seven per cent lower operating cost versus the competing A320 NEO.

John Hamilton went on the record stating that '...the 737 is a more efficient, lighter design and requires less thrust than other airplanes in this class, which is important because weight and thrust have a significant effect on fuel efficiency and operating costs. With airlines facing rising fuel costs and weight-based costs equating to nearly thirty per cent of an airline's operating costs, this optimised sixty-eight inch fan design will

The underside of TUIfly Boeing 737-8K5 D-AHFZ in CEWE Fotobuch Livery, demonstrating exactly why it was not possible to move the landing gear so as to be able to extend its length. *(author)*

offer a smaller, lighter and more fuel-efficient engine to ensure we maintain the current advantage we have over the competition'.

Airbus disputed this claim, with the spokeswoman at its North American unit Mary Anne Greczyn saying '...if a smaller fan engine were to generate the appropriate level of efficiency, we could have easily incorporated that, since we are not constrained as our competitor. The A320 NEO family is designed to benefit from the aircraft's inherent advantage'.

Boeing also opted to lengthen the nose landing gear of the 737 MAX to allow, in their words, better optimisation to take place. This was likely to necessitate the relocation of the narrowbody aircraft's electronics/equipment bay without a nose blister fairing. Boeing 737 chief programme engineer John Hamilton: 'We can put a sixty-eight-inch fan on the airplane without changing the nose gear but we allowed our designers to remove that constraint to see if they could further optimise the engine on the airplane and we believe there is a little better optimisation that will occur when we allow the nose gear to float up a little bit. Today the nose gear 737NG is actually slightly tilted down and so today's jetways, today's airstairs are not going to be affected by the change. We understand the nose gear design, and we will be finalising that in the months ahead'.

Boeing Commercial Airplanes president and chief executive Jim Albaugh: 'While the main driver of the ten-to-twelve per cent fuel burn saving remains the new CFM Leap-1B

Details and differences.

Above and below: Main and nose landing gear of a Southwest 737 MAX 8.

Right: the main gear of an Airbus A320 by comparison.

engines with a new strut that places the engine in a position much forward than the existing 737 NG, there will be design changes involving aerodynamics, software, the design of the winglet and more. For instance, the most distinctive aerodynamic change featured in the artists' renderings of the 737 MAX has been the adoption of a 787-styled tail cone, which will deliver better aerodynamics in the airflow through the empennage, thereby reducing drag of the airplane'.

Type Rating and MCAS

In the USA, the 737 MAX shares a common type rating with all the other Boeing 737 families. The impetus for Boeing to build the

737 MAX was serious competition from the Airbus A320 NEO, which was seen as a huge threat to win a major order for aircraft from American Airlines, a traditional customer for Boeing airliners. Boeing decided to update its 737, designed in the 1960s, rather than creating a brand-new airliner, which would have cost much more and taken years longer. Boeing's goal was to ensure the 737 MAX would not need a new type rating, which would require significant additional pilot training, adding unacceptably to the overall cost of the

Compare the underside of Air France's A320 November Oscar with that of the TUIfly Boeing 737-8K5 on an earlier page and the differences are very noticable. *(author)*

airliner for customers.

The 737 original and main certification was issued by the FAA in 1967. Like every new 737 model since then, the MAX has been approved partially with the original requirements and partially with more current regulations, enabling certain rules and requirements to be grandfathered in. 'Grandfathering' was a provision in which an old rule continues to apply to some existing situations while a new rule will apply to all future cases.

Chief executive Dai Whittingham of the independent trade group UK Flight Safety Committee disputed the idea that the MAX was just another 737, saying, 'It is a different body and aircraft but certifiers gave it the same type rating.'

Boeing decided that 737 pilots needed no extra training on the system - and indeed that they didn't even need to know about it. Boeing's stance allowed the new jet to earn a common type rating with existing 737 models, allowing airlines to minimize training of pilots moving to the MAX.

Much was said in 2019 in the mainstream media about what they described as 'a device' called MCAS that is fitted to the 737 MAX. They infer in somewhat hushed tones that this 'machine' can take-over control of the aircraft whenever it likes. It's not quite like that.

MCAS, the Maneuvering Control Augmentation System, is a flight control law - computer software if you like - built into the Boeing 737 MAX's flight control computer, designed to help the aircraft emulate the handling characteristics of the earlier Boeing 737NG. In simple terms, the software was designed and certified for the 737 MAX to enhance the pitch stability of the airliner so that it feels and flies like other 737s.

The software code for the MCAS function and the computer for executing the software are built to Boeing's specifications by Collins Aerospace, formerly Rockwell Collins.

According to an international Civil Aviation Authorities Team Review (JATR) commissioned by the FAA, MCAS may be a stall identification or protection system, depending on the natural stall characteristics of the aircraft. Boeing themselves considered MCAS part of the flight control system and elected not to describe it in the flight manual or training materials, based on the fundamental design philosophy of retaining commonality with the 737NG. MCAS was supposed to activate only in extreme circumstances far outside the normal flight envelope.

The 1,600-page flight crew manual mentioned the term MCAS once, in the glossary. Top Boeing officials believed MCAS operated only far beyond normal flight envelopes, that it was unlikely to activate in normal flight.

Dennis Tajer, a spokesman for the Allied Pilots Association at American Airlines, said his training on moving from the old 737 NG flight deck to the new 737 MAX consisted of little more than a one-hour session on an iPad, with no simulator training.

Minimizing MAX pilot transition training was an important cost saving for Boeing's airline customers, a key selling point for the airliner. Indeed, the company's website at the time pitched the jet to airlines with a promise that 'as you build your 737 MAX fleet, millions of dollars will be saved because of its commonality with the Next-Generation 737.'

When activated, MCAS directly engages the horizontal stabiliser, thus is distinct from an anti-stall device such as stick pusher, which physically moves the pilot's control column forward and engages the aircraft's elevators when the machine is approaching a stall.

Boeing CEO Dennis Muilenburg explained MCAS to the media in May 2019 saying that it '...has been reported or described as an anti-stall system, which it is not. It's a system that's designed to provide handling qualities for the pilot that meet pilot preferences.'

The 737 MAX's larger CFM LEAP-1B engines are fitted further forward and higher up than in previous models. The aerodynamic effect of its nacelles contributed to the aircraft's tendency to

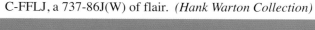

C-FFLJ, a 737-86J(W) of flair. *(Hank Warton Collection)*

With the Theme Building in the background, this could only be Los Angeles International Airport. This iconic space-age structure - used for much of its life as a restaurant - was influenced by 'Populuxe' architecture, and is an example of the Mid-century modern design movement later to become known as 'Googie'. The distinctive white building resembles a flying saucer that has landed on its four legs. The initial design was created by James Langenheim, of Pereira & Luckman, subsequently taken to fruition by a team of architects and engineers headed by William Pereira and Charles Luckman, that also included Paul Williams and Welton Becket. In the foreground, an Air Canada 737 launches skyward. *(Hank Warton Collection)*

pitch up at high Angles of Attack. The MCAS is intended to compensate in such cases, modelling the pitching behaviour of previous models, and meet a certain certification requirement, in order to enhance handling characteristics and thus minimizing the need for significant pilot retraining. Again, to simplify the explanation, MCAS was intended to modify the actual flying characteristics of the 737 MAX to mimic the behaviour of the previous variants that the pilots already knew.

As an automated corrective measure, the MCAS was given full authority to bring the aircraft nose down, and could not be overridden by pilot resistance against the control wheel as on previous versions of the 737.

The MCAS design parameters originally envisioned automated corrective actions to be taken in cases of high Angle of Attack and G-forces beyond normal flight conditions. Test pilots routinely push aircraft to such extremes, as the FAA requires aircraft to perform as expected. It was during such flights that test pilot Ray Craig determined the 737 MAX did not fly smoothly, in part due to the larger engines. Craig is on the record for stating that he would have preferred an aerodynamic solution. Still, Boeing decided to implement a control law in the software following suggestions from engineers who had worked on the KC-46A Pegasus, an aerial refuelling tanker that is fundamentally a Boeing 767-2C and which included the use of an MCAS function. The Pegasus team suggested MCAS to the MAX design team as a way of solving the problem.

This was, however, a slightly different version, The system took input on dual

The 737 flightline at Renton in early 2019. Commercial airliner skin panels arrive at Boeing covered with a temporary protective coating to protect the metal from damage or corrosion during the manufacturing and assembly processes. This is where the term 'green airplane' comes from; it references the green protective coating and implies that the product is unfinished, awaiting the paint shop's handiwork.
Paint shop employees first use a hand-sander on a completed airliner's previously primed surfaces, then tackle the green skins. They then mask off areas that need protecting, such as landing gear and engines, and apply a detergent to the aircraft to remove the coating. Next, they wash the aircraft with a fire hose spraying water heated to 100 degrees Fahrenheit to remove the green colour. This leaves the natural, silver-coloured aluminium skins. Once the jet is clean, painters sand or abrade the surface of the panels to ensure the primer coat adheres to the metal. They apply primer and then mask off areas of the jet and paint it in various stages, depending on the process for each customer's livery. They apply hand-detailing or decals last, prior to preparing the airliner for delivery to the customer. The time needed to complete a paint job depends on the size of the aircraft and the complexity of the livery.

redundant angle of attack sensors and would disengage with stick input by the pilot. The US Air Force stated that 'The KC-46 has protections that ensure pilot manual inputs have override priority' and that it 'does not fly the models of aircraft involved in the recent accidents' and that it is 'reviewing our procedures and training as part of our normal and ongoing review process.'

With the MCAS implemented, new test pilot Ed Wilson recorded that the 'MAX wasn't handling well when nearing stalls at low speeds' and recommended that the MCAS was applied across a broader range of flight conditions. Now MCAS was required to function under normal G-forces and, at stalling speeds, deflect the vertical trim more rapidly and to a greater extent. The original Boeing document provided to the FAA included a description specifying a limit to how much the system could move the horizontal tail - a limit of 0.6 degrees, out of a physical maximum of just less than 5 degrees of nose-down movement.

That limit was later increased after flight tests showed that a more powerful movement of the tail was required to avert a stall when the plane is in danger of losing lift and spiralling down.

MCAS also took its input data from a single Angle of Attack sensor close to the aircraft's nose despite a pair being available. This 'single point of failure' went against aviation requirements of robustness and integrity, for example using redundancy.

The FAA did not conduct a safety analysis on the changes, for it had already approved the previous version of MCAS, and the agency's rules did not require it to take a second look because the changes did not affect how the airliner operated in extreme situations.

Then came the Lion Air MAX 8 crash on 29 October 2018. Boeing acted reasonably quickly, issuing an Operations Manual Bulletin on 7 November which outlined the many indications and effects resulting from erroneous Angle of Attack data and provided instructions to turn off the motorised trim system for the remainder of the flight and trim manually instead.

The Angle of Attack sensor measures the angular difference between the direction the aircraft is moving and the pitch of the aircraft's wing. Until Boeing supplemented the manuals and training, pilots were unaware of the existence of MCAS due to its omission from the crew manual and no coverage in training. Boeing's bulletin to the airlines stated that the limit of MCAS's command was 2.5 degrees. That number was new to FAA engineers who had seen 0.6 degrees in the original safety assessment. 'The FAA believed the airliner was designed to the 0.6 limit, and that's what the foreign regulatory authorities thought, too' said one FAA engineer speaking under a guarentee of anonymity; '...It makes a difference in your assessment of the hazard involved'.

The higher limit meant that each time MCAS was triggered, it caused a much greater movement of the tail than was specified in that original safety analysis document.

If the final safety analysis document was updated in parts, it certainly still contained the 0.6 limit in some places and the update was not widely communicated within the FAA technical evaluation team. 'None of the engineers were aware of a higher limit,' said a second current FAA engineer.

The discrepancy over this number was magnified by another element in the System Safety Analysis: the limit of the system's authority to move the tail applies each time MCAS is triggered. And it can be triggered multiple times, as it was on the Lion Air flight.

One FAA safety engineer said that every time the pilots on the Lion Air flight reset the switches on their control columns to pull the nose back up, MCAS would have kicked in again and 'allowed new increments of 2.5 degrees. So once they pushed a couple of times, they were at full stop,' meaning at the full extent of the tail swivel, he said.

The FAA followed the ops manual update with an emergency airworthiness directive against the aircraft, warning operators the pitch-over threat exists even

A number of Southwest Airlines 737 MAX-8s in storage. *(Hank Warton Collection)*

when the aircraft is being hand flown by pilots. The agency said operators had less than thirty days to comply with the directive.

This Airworthiness Directive brought out speculation in the media - and investigators were cautious not to claim the Angle of Attack issue addressed by the directive as the only cause of the accident.

Concerns focused on both the Angle of Attack indicator itself and the software that transmits the indicator's information to the aircraft. There were concerns as to whether incorrect data from the Angle of Attack sensor being fed to other aircraft systems may have caused the aircraft to pitch over on its own or whether the cockpit crew may have misinterpreted potentially erroneous flight instrument indications.

Eventually, evaluation of the data recorders on Lion Air 610 revealed that when the MCAS pushed the jet's nose down, the captain pulled it back up, using thumb switches on the control column. Still operating under the false Angle-of-Attack reading, MCAS kicked in each time to swivel the horizontal tail and push the nose down again.

The black box data showed that after this cycle was repeated twenty-one times, the Captain ceded control to the First Officer. As MCAS pushed the nose down two or three times more, the first officer responded with only two short flicks of the thumb switches.

At a limit of 2.5 degrees, two cycles of MCAS without correction would have been enough to reach the maximum nose-down effect.

In the final seconds, the black box data showed the Captain resumed control and pulled back up with high force. But it was too late. The airliner dived into the sea at more than 500 miles per hour.

After the Lion Air crash, 737 MAX pilots worldwide were notified about the existence of MCAS and what to do if the system is triggered inappropriately.

Boeing insisted that the pilots on the Lion Air flight should have recognized that the horizontal stabiliser was moving uncommanded, and should have responded with a standard pilot checklist procedure to handle what's called 'stabiliser runaway.' If they'd done so, the pilots would have hit cutoff switches and deactivated the automatic stabiliser movement. Boeing pointed out that the pilots flying the same

United's 737-MAX9 N37504, seen at Los Angeles. *(Hank Warton Collection)*

A very striking scheme worn by 9M-AAI a 737-301 of Air Asia, leased from US Airways.
(Wan Hunglo Collection)

aircraft on the day before the crash experienced similar behaviour to Flight 610 and did exactly that: they threw the stabiliser cutoff switches, regained control and continued with the rest of the flight.

However, what happened on the Lion Air flight 610 did look like a standard stabiliser runaway, because that is defined as continuous uncommanded movement of the tail. On the accident flight, the tail movement wasn't continuous; the pilots were able to counter the nose-down movement multiple times.

In addition, the MCAS altered the control column response to the stabiliser movement. Pulling back on the column normally interrupts any stabiliser nose-down movement, but with MCAS operating that control column function was disabled.

These differences certainly could have confused the Lion Air pilots as to what was going on.

Boeing's safety analysis of the system assumed that the pilots would recognize what was happening as a runaway and cut off the switches. As one engineer put it: 'The assumptions are incorrect. The human factors were not properly evaluated'.

As was expected, Murphy's Law - that of anything that can go wrong, will go wrong - kicked in during the fixes and sure enough another problem was revealed. It was at least the third different software problem to be discovered since the airliner was grounded in March 2019.

The new issue apparently had to do with a warning light that helped tell pilots when the trim system - a part of the aircraft that could raise or lower the nose - was not working correctly. Federal Aviation Administration head Steve Dickson said during a talk in London that the light was 'staying on for longer than a desired period.'

Boeing and the FAA had previously disclosed two other glitches that were discovered during the top-to-bottom review of the MAX. In January, Boeing announced that it found a problem in the startup process of the airliner's flight computers, which was serious enough for the company and the FAA to delay a crucial test flight. That followed a previous flaw in the flight computer discovered in June 2019 that the FAA said 'could cause the plane to dive in a way that pilots had difficulty recovering from in simulator tests.'

The new glitch was a direct result of the fixes Boeing made to those previous flaws. The trim system flaw 'resulted from the redesign of the two flight computers that

flydubai's 737-MAX8 A6-FMB seen just about to pushback in Arctic conditions before the grounding. *(Ameen Yasheed collection)*

control the 737 Max to make them more resilient to failure. The change has been incorporated to the 737 Max'.

February 2020 - the discoveries keep occurring.

14 February and it came to light that during the original design and certification of the MAX, company engineers failed to notice that the electrical wiring did not meet the current regulations for safe wire separation - something that the FAA also missed.

Boeing discovered the wiring vulnerabilities and told the FAA of the problem when, after the crashes, it undertook a complete redo of its system safety analysis on the MAX. It was forced to do the new analysis when it realised the MAX's original certification analysis included assumptions about pilot reaction times that did not match the reality of the responses during the two MAX crashes.

It was unclear how during the design of the MAX Boeing missed the fact that the wiring didn't meet the regulation governing the separation of wires to prevent short-circuits. The regulation was introduced in 2009 following studies of two fatal crashes: TWA Flight 800 in 1996, in which an electrical short-circuit was believed to have caused a spark in the fuel tank and an explosion in Boeing 747 N93119; and Swissair 111 in 1998, when an electrical short-circuit caused a fire in the cockpit in a McDonnell Douglas MD-11 HB-IWF.

The FAA safety engineer said Boeing identified about a dozen positions in the 737 wiring loom, including one toward the airliner's tail and the rest in the electronics bay under the forward fuselage, where significant runs of wire failed to meet the new separation standard. The wire lengths involved were as long as sixteen feet.

In one instance, engineers found a hot power wire too close to two command wires running to the aircraft's moveable stabiliser, one for commanding the tail to swivel to move the jet nose-up, the other to move it nose-down. The danger was a short circuit that would cause arcing of electricity from the hot wire to the command wire. If a hot short occurred between the power wire and either the up or down command wire, the stabiliser could go to the full nose-up or nose-down position.

Furthermore, the electrical power in that wire could circumvent the cutoff switches in the cockpit that, in the event of such a stabiliser runaway, are used to kill electrical power to the tail. Theoretically, the pilots could be unable to shut it off.

The arcing danger was considered to be remote, but was scientifically established. Michael Traskos, chairman of the industry's wiring and cable standards committee and president of Lectromec, a Virginia-based laboratory and engineering firm specialising in wire-system component testing and consulting, said that his team did testing for NASA in 2005 and 2006, not specific to a particular airliner, but that their tests had 'demonstrated potential uncommanded activation in the event of arcing.'

The wiring vulnerability created the

theoretical potential for an electrical short of moving the airliner's horizontal tail uncommanded by the pilot, which could be catastrophic. If that were to happen, it could lead to a flight control emergency similar to the one that brought down two MAXs.

The danger was considered as being remote, but the FAA faced a dilemma as to what to do about it. Modifying the wiring would be a delicate and expensive task, and in early February 2020 Boeing submitted to the FAA that it should not be required.

Boeing's argument rested on the service history of the earlier 737NG, which had the same wiring, but did not have to meet the current wiring-separation standards because they came into force long after that jet was certified. A Boeing official claimed in a statement 'There are 205 million flight hours in the 737 fleet with this wiring type. There have been sixteen failures in service, none of which were applicable to this scenario. We've had no hot shorts.' In addition, Boeing said pulling out and rerouting wires on the MAXs already built would pose a higher risk of causing an electrical short because insulation could chafe or crack in the process of moving the wires.

Yet allowing the wiring to remain as it was would be difficult at a time when both Boeing and the FAA were under tremendous scrutiny. One FAA safety engineer said that agency technical staff had been clear that the wiring did not comply with regulations and had told their Boeing counterparts it had to be fixed.

Another person familiar with the FAA's thinking said the agency had communicated to Boeing that despite the safe service history of the wiring on other 737s, it would be difficult to convince regulators that they should do nothing. 'Our people [the FAA] have to weigh that against the regulations and the political and public opinion risk of appearing to give Boeing a break on a regulation that's there for a reason'. There was also intense pressure from overseas regulators, including the European Union Aviation Safety Agency (EASA). 'It's probably true that if Boeing proposes to do nothing, EASA is going to say, 'Hell, no'.

On 14 February 2020 the FAA issued an official statement hinting that Boeing could be forced to comply with the wiring regulation. 'We will rigorously evaluate Boeing's proposal to address a recently discovered wiring issue with the 737 MAX. The manufacturer must demonstrate

Another 'green' aircraft, this time a MAX 8 for American airlines if the rudder colours are anything to go by.*(BJ)*

compliance with all certification standards.'

Only a matter of a few days later - on 19 February - it was revealed that Boeing had discovered debris left inside the wing fuel tanks of several undelivered 737 MAXs during the aircraft assembly process. This was the latest of a string of quality control issues across three product lines. With the information becoming public on 19 February, Boeing was forced to order inspections of all the undelivered MAXs stored at various locations.

Inspections were also ordered on the 385 MAXs already delivered to customers, which had been grounded and were parked at airfields around the world. Company spokesman Bernard Choi said Boeing was recommending inspections for these airliners. 'It's still undecided if we will inspect the rest of the 737 fleets. Obviously, we'll do what's right for safety.'

Mark Jenks, vice president and general manager of the 737 programme and its Renton assembly site, sent all 737 employees a message outlining a series of actions to deal with the problem. 'Foreign Object Debris,' or FOD, is absolutely unacceptable. We need our entire team to make this a priority.'

The message to employees did not specify what debris had been found inside the wings, saying only that FOD had been recently found in the fuel tanks of several 737 MAX airliners in storage. However, it later came to light that the objects found included left-behind tools and rags. Either would be cause for serious concern. Tools banging around inside a fuel tank could damage sensors or wiring. A rag could block a fuel line. The new FOD inspections were expected to take two to three days per aircraft. Fuel had be drained from the wings before a mechanic could go in and do a thorough check.

Clearly there was an ongoing problem with FOD discoveries in the past couple of years. Previously on the 767-based KC-46 military tanker built in Everett and on the 787 Dreamliners built in North Charleston, SC there was a systemic issue with Boeing's quality control that had not been fully contained.

In 2019 Boeing was forced to ground its KC-46 tankers after the Air Force expressed concern about loose tools and bits of debris found in various locations inside the completed airframes. Management in Everett declared a level 3 state of alert on the assembly line over the KC-46 FOD issue, which was a defence contract term just one step away from a complete shutdown of the assembly line. A memo sent to employees at that time urged them to win back the confidence of the Air Force 'and show them that we are the number one aircraft builder.'

2019 also saw a number of media sources reporting shoddy quality control on the 787 Dreamliner assembly lines in South Carolina dating back to 2016. Debris was routinely found dangerously close to wiring beneath cockpits at the assembly plant. Piles of titanium swarf - chips of metal produced by the machining process - had accumulated close to electrical equipment underneath the passenger floor. Employees found tubes of sealant and metal nuts inside finished jets. In one instance, a ladder and a string of lights were left inside the tail of a 787.

Less than a month after the crash of the second 737 MAX jet, Boeing called North Charleston employees to an urgent meeting about airlines finding random objects in their new airliners.

These lapses in FOD control came after Boeing announced a transformation of its manufacturing procedures that would result in it cutting almost 1,000 quality inspectors' jobs over two years.

The stready drip of bad news continued. The FAA's Transport Airplane Directorate memo to FAA Aircraft Certification Service on the subject of Lightning Protection for

A close-up of the starboard LEAP-1B engine attached to N87040, also known as 1A004, the fourth test machine, seen at Farnborough in 2016. *(author)*

Boeing 787 Fuel Tanks as described earlier struck the 737 MAX on 26 February 2020 when it was revealed that the FAA had prepared an airworthiness directive requiring all Boeing 737 MAXs to be inspected for a manufacturing defect that Boeing had discovered in December 2019 that harked back to machines manufactured between February 2018 and June 2019.

The defect arose when mechanics working on the final finish of the airliners, polishing the carbon composite engine nacelles at the end of the production process, ground away underlying layers of metal foil in the upper part of the nacelle serving as a shield against a surge of electrical current needed for lightning protection.

Without it lightning could 'induce spurious signals onto the underlying airplane wiring, including wiring associated with the engine control systems,' which could cause a loss of thrust control to one or both engines which could potentially lead to a dual engine power loss event. If not addressed, the condition could result in a forced landing away from an airport due to loss of thrust control on both engines'.

The fix required replacement of two carbon composite fairing panels that cover the area where the nacelle connects to the wing. Sealant to establish a required electrical bond path to safely disperse any lightning strike would also have to be applied. The FAA estimated that the airworthiness directive would affect 128 of Boeing's 737 MAX aircraft.

Another month, another problem.

9 March 2020 and a Southwest Boeing 737-7H4(W), registration N726SW, which first flew on 11 February 1999 according to records, was operating as Flight WN-1685 from Las Vegas, Nevada to Boise, Idaho. The airliner was en-route at FL390 about 130 nautical miles southwest of Salt Lake City, Utah when the crew initiated a rapid descent to FL220 due to unstable cabin pressure. The aircraft levelled off at FL220 about seven minutes after leaving FL390. The cabin pressure stabilized at FL220, and the passenger oxygen masks had not deployed so that the crew continued the flight to Boise maintaining FL220. The aircraft landed safely in Boise about 38 minutes after leaving FL390.

Then, on 13 March, the FAA reported the cabin pressure had gradually decreased in flight. An inspection by the FAA revealed a twelve-inch long crack in the crown skin of the aircraft in an area behind the flight deck, which was required to be inspected every 1500 hours. The last inspection of that

El Al's 4X-EKM, a 737-804(W) comes in for a landing at Malaga, Spain after a flight from Tel Aviv. *(Ameen Yasheed collection)*

area had been done 500 flight hours before the occurrence. Southwest spokeswoman Michelle Agnew said the airliner had been inspected for cracks within the required 1,500-flight period.

Agnew said that during the flight, pilots responded to an indicator in the cockpit and 'followed standard procedures by descending to a lower altitude to maintain a safe and comfortable cabin environment.' That 'resolved the issue,' and the crew continued safely to Boise.

'The aircraft did not incur a rapid depressurization, masks were not deployed, and the aircraft did not require a diversion to maintain safety of flight.'

A spokesman for Boeing said the manufacturer was aware of the incident and was working with Southwest to learn more. Apparently Southwest had told the FAA that required inspections had located external cracks in two other machines in the same area but that those incidents did not lead to cabin decompression. It brought back memories of Aloha Flight 243.

The implications of this were significant. The aircraft in question was 'old', just over twenty-one years since manufacture, but if maintained correctly, it certainly did not fall into the 'geriatric jets' category. At least three machines suffered from the same fault - all allegedly having gone through inspection at the supposedly correct time cycles. Was this an indication that there was another fleet-wide problem lurking?

What needs to be done to return to flight?

Return to flight depends on a number of aspects:

- Approval of the MCAS software 'fix' by the FAA and approval of any the modifications required to each airframe.
- Updating the computers onboard each aircraft and addition of extra sensors.
- Additional training of flight deck crews as to MCAS as the cause and effect thereof. This will require an as-yet-unspecified amount of simulator time.
- Convincing airline staff and the public that the MAX is safe to fly in.

Each airliner will have to be brought up to flight status after the period of storage. At best this will entail cleaning. It is more likely to entail more detailed work, including changing of inhibiting fluids used to protect the structure as many have been stored in the cold and wet Washington State climate over the winter months.

As far as can be ascertained, each individual aircraft will have to be inspected by FAA personnel. There will be no blanket clearance. Indeed, in August 2020 the FAA put forward a wide-ranging list of changes it wants to be made before the airliner could fly again commercially. At the same time, the European Union Aviation Safety Agency (EASA) said its own tests would take place in the week beginning 7 September. EASA maintained that clearance by the US FAA would not automatically mean an approval to fly in Europe. Boeing hopes to get the 737 Max back in the air sometime in 2021.

With the aerospace manufacturing sector expecting Boeing to restart 737 MAX production, industry insiders were speculating that it would take up to two years to clear out the stored inventory of narrowbody aircraft and fuselages. This was due to the calculations that it would take 18-24 months to push out the roughly 400 undelivered MAXs stockpiled by Boeing, in addition to the 387 grounded MAXs at customers, along with the almost 100 fuselages that aerostructures supplier Spirit AeroSystems has parked in Wichita.

With Boeing's proposal that MAX pilots should undergo simulator training, that requirement would add two-to-five months to the push-out timeline. As of March 2020 there were only 36 known MAX simulators operating worldwide, although major provider CAE continues to build so-called white tail simulators for sale 'off the shelf'.

Once suppliers restart production, it could take up to six months until the first delivery of a new-built MAX aircraft. Boeing CEO and president David Calhoun and CFO Greg Smith cautioned stakeholders that they will build production slowly, as Boeing's empty production system reset from its first position.

'We had a very orderly shutdown of the line,' Smith told an investor conference on 12 February. 'When you look at those three lines, they're empty.'

'The needle we're threading is we've got airplanes that are out on the ramp. We've got to get those delivered. We've got our factory, our supply chain and then ultimately customers' ability to take the airplanes. And then, of course, we've got the added element in that delivery now with the FAA being involved in every delivery of every MAX airplane that we've assumed will continue.'

Clearly Boeing has a lot to coordinate when it restarts the line. With all of that factored in, it means the MAX ramp up will be gradual, with deliveries likely starting in the fourth quarter, which begun in October.

One US survey in late 2019 revealed that seventy per cent of the flying public has reservations about flying in the 737 MAX once it was cleared to return to the skies.

Boeing started to prepare for exactly this scenario by producing and sending out to airlines a draft copy of what it called a 'Customer journey and scenario map' in the style of a spreadsheet. In this document, various stages of the customer's journey was broken down into different elements: Booking a Trip, Trip Booked and Checking In in advance, At the Gate, On the Plane (Prior to Takeoff) and On the Plane (Inflight). The next column gave the perceived mindset of the passenger - all of them gave just one - 'anxious' when the passenger realised they were about to or were flying on a 737 MAX.

Various scenarios were envisaged in the next column: at booking or prior to checking in the passengers realise they are about to fly on a MAX and posts publically on social media criticism of the airline using the MAX or their fear of flying on the airliner. Then, at the gate or onboard the aircraft, the passenger panics creates a scene that causes anxiety for others and is captured on video and is shared on social media.

The next column - by far the largest - provided details of what Boeing thought were the ways for the airlines to react, under the heading of 'Possible Airliner Passenger Support'. If the passenger expressed criticism of the airline for using the MAX at the booking the flight stage or when doing check-in in advance, Boeing suggested that

The scimitar winglet as fitted to the 737 MAX-8 *(author)*

the airline proactively engaged the passenger on social media to provide information on the safety of the MAX, and then offer to take the conversation offline to discuss alternative travel arrangements if the passenger was still anxious.

If the passenger panics at the gate, ramp agents are to offer information as to the safety of the MAX, and show a willingness to re-book an alternative flight. If needed, the ramp agent should seek pilot intervention to inspire the passenger with confidence. Airline social care teams should link in with any social media postings, and through the comments threads express concern for the passenger's anxiety. The social care teams should also offer to re-book.

A similar process was advised if the passenger panicked when onboard, but still on the ground. A Flight Attendant could offer a de-board and re-book, or call for pilot intervention. If the anxious passenger starts posting on social media, the airline social care team was to provide information on the safety of the MAX and then inform the airlines Network Operations Center of the anxious passenger, which informs the flight crew as appropriate.

Finally, there was a section regarding any in-flight incident. The Flight Attendant should approach the customer to offer comfort and information on the safety of the MAX. The flight crew should use techniques related to an inflight medical emergency to de-escalate the situation. Dealing with social media contact was to be the same as previous.

Boeing suggested a whole series of items to be available called 'toolkit materials' for airline staff; Boeing Pilot videos, Pilot Confidence videos, Boeing Explainer videos, Explainer storyboards, 3x5 cards and FAQ cards for frontline employees, 3x5 cards and FAQ cards for pilots, FAQ cards for passengers and key point cards for frontline employees, pilots and passengers.

Chapter Fifteen

Can Boeing survive the 737?

For decades, America's leading aircraft maker was so revered by flyers and employees alike that it inspired a bumper sticker and a huge billboard on the escarpment between Mukilteo and Paine Field: 'If it's not Boeing, I'm not going.' By early 2020 in the minds of many, one word had been deleted; 'If it's Boeing, I'm not going.'

It was not surprising that it did not take long for the 'armchair experts' to surface and start making their opinions known on social media. Much of the comments were garbage, but there were meaningful insights buried in the rubble of mangled facts.

Expert A: 'For the past decade, the greatest priority of Boeing's top management has been increasing their share price - not via making incredible products, but by investing major resources into stock buy-backs. And so, they cut union jobs, laid off quality inspectors, etc - but, hey, the stock price went up. And, now that the consequences of their greed has become clear, the parasites will depart to find other sources of easy money while the wreckage of a once-great aerospace company struggles to survive on what little reputation they have left.'

Expert B: 'Boeing needs to close the Chicago office, fire all the sycophant MBA's and move executives into the production floor to find out WTF is going on. These monkeys put an extremely valuable company on the skids while stuffing their pie holes with cash. It is criminal'.

Expert C: 'Looks like the outsource-everything, then just snap it all together approach doesn't scale up from iPhones to aircraft. No worries though, at least not for the management - the CEO who signed off on the MAX back in 2011 retired with a $300k pension. Per month. The CEO who was in charge during the two crashes that killed 346 was 'fired', i.e., given $62.2 million to go away. Nikki Haley remains on Boeing's board collecting $300k/yr for attending a few meetings. And shareholders made out great, milking Boeing for $78 billion over the last 15 years. Only the little people, the ones on the factory floor, stand to lose their livelihoods, and most likely their homes. Ain't free enterprise great?

A short comment from one observer summed it all up: 'I'll take an Airbus over a Boeing any time. That's Boeing's fault'.

Boeing's problems with the 737 MAX throughout 2019 fanned speculation that travellers could well fear flying aboard the airliner even when regulators lift the order grounding it. In early 2020 that day was probably months away as Boeing stopped building the jet. Nevertheless, the aerospace giant remained hobbled by continued revelations, including internal memos exposing a culture focused on evading regulators at the cost of safety and reliability.

In December 2019, when Boeing's board of directors fired CEO Dennis Muilenburg, some corporate governance experts and investment analysts wondered why it took so long.

A blast from the past! G-FIGP, the former G-BMDF, a 737-2E7 of Dan-Air, is seen here in the European Skybus markings to be operated by the Paul Stoddard Minardi F1 racing team. *(author)*

By then, the 737 MAX, had been grounded worldwide for more than nine months. Investigations, lawsuits and news stories revealing festering internal problems had piled up. Muilenburg faced public grillings and calls for his resignation during Congressional hearings. Revenues and stock values plummeted. Worst of all, Boeing's once sterling reputation for quality and safety had been badly tarnished.

The hesitancy to fire Muilenberg in the face of spiralling crises underscored concerns about the board's oversight of the company, even as it faced the most troubling period of its 103-year history. It also raised questions about the board's guilt in the tragedies and its ability to re-establish confidence in the company among regulators, Wall Street and the flying public.

It was not until Muilenburg backpedalled from his steadfast forecast that the MAX would be cleared to fly again by the close of 2019 that the board's confidence in him finally broke. Once the timeline for the MAX's recertification fell apart, the company's already strained relationships with suppliers, airline customers and regulators reached a tipping point.

David Calhoun, a Boeing director for a decade and Muilenburg's eventual replacement, later told reporters '..every board meeting ended with a confidence question, and we always asked the question: Does he have it? We stayed with it and we stayed with it, and then we got into December, where we really did have sort of a schedule that was committed to in a really big way. The vote at that moment in time, at the end of the last meeting, simply was the end of our confidence' Calhoun said of the board's 22 December decision. 'And then, we moved forward'.

Calhoun claimed board members didn't learn of the various problems that ultimately surfaced during Boeing's developing crisis until too late in the game.

The Boeing board boasted big-name officials with varied backgrounds in business and industry, government and the military. They included three former US ambassadors (Caroline Kennedy, Nikki Haley and Susan Schwab); two retired Navy admirals (Edmund Giambastiani Jr. and John Richardson); and a slate of current or former corporate executives with experience in medical and biotech industries (Robert Bradway and Arthur Collins Jr.); electrical utilities (Lynn Good); air carriers (Larry

Kellner); insurance companies (Edward Liddy and Ronald Williams) and private equity (Mike Zafirovski).

Calhoun, himself a board member since 2009, held executive roles at the Blackstone Group private equity firm, marketing researcher Nielsen and General Electric.

Corporate boards, which exist to ensure publicly traded companies are adequately managed for shareholders, seek a mix of directors that balance different competencies. The bulk of these boards will be current and former CEOs of big companies, then maybe someone with international trade expertise, and a senator with connections in Washington, DC. There might also be a high profile Nobel Prize winner or another prominent name that looks good to the outside world.

Boeing's board, while not lacking in name recognition or business prowess, has drawn criticism for an apparent deficit of technical, safety and engineering expertise. It seems that this was a byproduct of Jim McNerney, a longtime GE administrator who, in 2005, became Boeing's first chief executive without an aviation background. The Boeing board of early 2020 was largely created by McNerney, and he packed it with people with no engineering experience.

Into the crisis stepped a new leader: David Calhoun, who started as CEO on 13 January. Calhoun set his sights firmly on fixing the 737 MAX. One of his first tasks was to send an email to employees. 'This must be our primary focus. This includes following the lead of our regulators and working with them to ensure they're completely satisfied with the airplane and our work.'

It did not take Calhoun long to come to one overriding conclusion: Things inside the aerospace giant were even worse than he had thought. Calhoun criticised his predecessor in blunt terms and went on record saying that he was focused on transforming the internal culture of a company mired in crisis.

To get Boeing back on track Calhoun said he was working to mend relationships with angry airlines, win back the confidence of international regulators and appease an anxious President Trump - all while moving as quickly as possible to get the grounded 737 MAX back in the air.

'It's more than I imagined it would be, honestly and it speaks to the weaknesses of our leadership'.

The 737 MAX was only the start. These days, Boeing, whose history included such legendary machines as the B-17 'Flying Fortress' bomber and 747 jumbo jet, seemed to suffer only setbacks. In December 2019, Boeing's Starliner astronaut capsule failed to reach the proper orbit on an unmanned test flight.

An interim assessment of what went wrong during December's first uncrewed flight of Boeing's CST-100 Starliner space taxi turned up so many breakdowns that NASA ordered a comprehensive safety review of the company's procedures.

NASA and Boeing provided a status report on the Starliner post-flight reviews on 8 February 2020, after concerns were raised publicly during a meeting of the space agency's Aerospace Safety Advisory Panel.

'We are 100 per cent committed to transparency,' NASA Administrator Jim Bridenstine told reporters during a teleconference.

NASA said that the revelations added to concerns about engineering shortcomings in other lines of Boeing's business - including commercial airliners, where a software issue and lapses in training procedures led to problems with 737 MAXs; and military aircraft, where Boeing was having to retrofit Air Force KC-46 tankers to fix a design flaw. Douglas Loverro, NASA's associate administrator for human exploration and operations, alluded to those shortcomings as he discussed the decision he made with Bridenstine's support to order a wider safety review. 'There were several factors that were in my mind when I asked the boss if we could do this,' he said. 'And those were obviously press reports that we've seen from other parts of Boeing, as well as what seemed to be characterized as these software issues.' It seems that NASA had started to notice a pattern.

Boeing's Orbital Test Flight fell short of its goal not long after its 20 December launch when the rocket engine failed to light up on schedule. By the time engineers could fix the problem and upload the corrected software, it was too late to get the spacecraft on track for its planned rendezvous with the International Space Station.

During their troubleshooting, engineers found that a programme designed to keep track of the mission elapsed time - that is, how much time had elapsed since launch - had picked up a time stamp from the United Launch Alliance Atlas 5 booster almost 11 hours too early and never corrected the time. That issue alone threw the schedule for sending three space fliers on a follow-up mission to the station into limbo. But there was another software issue that didn't become public until the safety panel meeting in early February.

Boeing and NASA officials reported that the issue could have ruined the separation of the Starliner's service module from its crew module during descent. Jim Chilton, senior vice president of Boeing Space and Launch, said that the service module's thrusters were programmed to follow the wrong firing profile, with the result that the module might not have achieved enough separation. 'That could have resulted in the service module bumping into the crew module,' Chilton said. NASA said such a scenario could have resulted in the loss of the vehicle. Chilton said Boeing engineers hadn't fully worked through what would have happened. However, he added, 'it can't be good when two spacecraft are going to contact.'

Somehow, both of the software issues had slipped through pre-launch verification and testing of Starliner's flight software. Chilton said the second problem was discovered on the night before the spacecraft was to land, during a comprehensive double-check of the software. 'We found it because we went looking,' Chilton said.

Engineers rushed to rewrite the software, verified that it ran correctly, and uploaded the fix less than three hours before the spacecraft's 22 December landing, said John Mulholland, vice president and program manager for Boeing's Starliner programme. Thanks to the fix, the Starliner made a trouble-free touchdown in New Mexico.

Yet another problem cropped up during the flight: Starliner's communication system had a hard time establishing contact with NASA's relay satellites, which contributed to the delay in uploading the fix for the initial engine-firing problem.

On the day after launch, Chilton said the problem cropped up because Starliner's misfiring thrusters put it in the wrong position to make contact with the satellite network. Mulholland reported that Boeing's preliminary conclusion was that the problem had to do with 'Earth-generated noise over specific geographic footprints.'

'We believe the frequency ... is very close to the frequency that would be associated with cellphone towers, and that created that high noise floor which didn't allow us to establish a link as soon as we would have needed to,' he said.

He stressed that the communication issue was still under investigation. A joint Boeing-NASA investigative team was due to finish its work by the end of February.

Loverro said he was concerned that the Starliner programme's problems could go deeper than the three issues that cropped up during December's mission.

'To put it bluntly, the issue that we're dealing with is that we have numerous process escapes in the software design, development and test cycle for Starliner,' he said. 'The two software issues that you all know about are indicators of the software problems, but they are likely only symptoms. They are not the real problem.'

Loverro compared the job to check a car's spare tyre. 'If you have a flat spare tire, you don't find that out by driving the car,' he said. 'You find it out by opening the trunk and checking the pressure in the tire. ... Our intention is to go check the spare and make sure it's inflated properly.'

Kathy Lueders, programme manager for NASA's Commercial Crew Program, said the investigative team had already come up with a list of eleven high-priority corrective actions, mostly having to do with addressing known software errors and tightening up testing procedures.

'We're going to take those actions, along with further actions that Doug and the EO (Exploration and Operations) team have given us, to make sure that we're really addressing root cause,' she said.

One of the requirements was to take a closer look at the full Starliner software package, which Boeing said amounts to a million lines of code. Bridenstine said it was too early to predict how the safety review might affect plans for future Starliner flights.

That capped a year of embarrassments, including the Air Force's decision to suspend deliveries of the new KC-46 air tanker after finding manufacturing debris that had been left aboard.

Those issues pale compared with the nightmare that is the 737 MAX. Boeing kept offering rosy predictions for its imminent return even as airlines kept pushing back the date on their schedules.

Jim Hall, former chairman of the National Transportation Safety Board: 'It's hard to put a time frame on it. If it requires (pilot) retraining or more extensive evaluation around the world, it could be the end of the year.'

Others had given up making predictions. 'We put our calendars in the drawer and took our watches off' said Dennis Tajer, spokesman for American Airlines' pilot union.

In 2019 customers cancelled more orders for Boeing's airliners than they placed, in part because airlines weren't clamoring to order the grounded 737 MAX.

Calhoun urged a fresh start for the company. He wanted his staff to better listen to customers. 'Every day, we will commit to our shared values while further

Love or loathe the CP Air partial bare-metal scheme, the sun sparkles on C-FCPG, a 737-317. *(Matt Black Collection)*

strengthening our culture,' he told employees.

He admitted that Boeing's reputation was hurt by the crisis, and by some of the internal company communications made public. 'My stomach turned, just like yours would, just like anyone's would,' he said when asked his reaction when he first read the communications.

Calhoun reiterated that despite some suggestions that it would, the company had no plans to change the name of the airliner. 'I'm not going to market my way out of this.'

It was a very laudable intention, but I am not sure that the company will have any control in the matter. If the travelling public has lost faith in the product, then they will show a reticence to travel on it, and that will be something that the airlines' marketing teams will keep a very close eye on. At the first sign of reluctance, the marketing people will report back to the manufacturer and pile on the pressure to do something - and do something fast.

One of Calhoun's first tasks as chief executive was to go on an apology tour, holding a series of what he called 'greet-and-mend opportunities'. The first stop was the White House after just three days in the job. At a private meeting with Donald Trump, the president told him that he liked Dennis Muilenburg but believed a leadership change had been needed. The president said he hoped Boeing was investing all of its resources into getting the plane back in the air. 'He wants us to get back on our horse. He wants us to get the MAX flying again, safely'.

Calhoun also went on record saying that he was confident that the MAX would eventually be able to return to service safely. 'If we didn't believe we were going to field a plane that is safer than the safest demonstrated aircraft that is out there today, we wouldn't do it' he said in one interview. He also reiterated that airline customers remained supportive of Boeing, although he did admit 'I would never describe our discussions today as cordial. They just want us to get back on track.'

Boeing's 29 January online and telephone conference with media and investor attendees will surely be remembered as one of the most humbling earnings reports in the 103-year-old company's history. For 737 MAX suppliers, it set the tone for the next two or three years.

'Slowly' was the word repeated ad nauseum by Boeing leadership, especially when it came to the narrowbody's monthly production rate ramp-up - and that assumed the MAX was set to become 'ungrounded' and production resumed. Boeing CEO David Calhoun and Chief Financial Officer Greg Smith said during the conference that production flow will return 'one step at a time', even 'one airplane at a time'.

Boeing would not reveal production plans beyond stating it could restart manufacturing months before the MAX was returned to service. But major suppliers, analysts and consultants were piecing together a road map that foresaw MAX monthly unit production rates hitting the mid-twenties in 2020, the thirties for most of 2021 and possibly back to fifty-two - where it stood before the MAX crisis - by the end of 2022.

Boeing indicated a new public marker of a midyear MAX return to service that was seemingly backed by the FAA, suggesting that the supply chain was expecting Boeing's own production to restart in March or April. 'We've assumed roughly a 90-day production delay, which is consistent with the direction that we've received from Boeing,' says United Technologies Corp. (UTC) Chairman and CEO Greg Hayes.

More anecdotal evidence came a day later from aerostructures leader Spirit AeroSystems, when managers announced a

9Y-TAB, a 737-8Q8 of BWIA West Indies. BWIA West Indies Airways Limited, known locally as 'Bee-Wee' and also as British West Indian Airways, was the national airline based in Trinidad and Tobago. At the end of operations on 31 December 2006 BWIA was the largest airline operating out of the Caribbean, with direct service to the United States, Canada, and the United Kingdom. Its main base was Piarco International Airport , Piarco, with major hubs at Grantley Adams International Airport and Cheddi Jagan International Airport. It was headquartered in the BWIA Administration Building in Piarco, Tunapuna–Piarco on the island of Trinidad. *(Kirk Smeeton Collection)*

new agreement with Boeing over MAX production. 'Under the agreement, Spirit will restart production slowly, ramping up deliveries throughout the year to reach a total of 216 MAX shipsets delivered to Boeing in 2020. Spirit does not expect to achieve a production rate of fifty-two shipsets per month until late 2022'.

That implied an average production rate of twenty-four a month in 2020, according to financial analysts Sheila Kahyaoglu and Greg Konrad at Jefferies, who noted that Spirit already counted about ninety-five 737 units parked in or near its factory.

Other suppliers were making similar noises. Hayes of UTC, home to Collins Aerospace and Pratt & Whitney, said managers assumed an average production rate of twenty-one airframes per month in the second half of the year for Collins. General Electric executives said their LEAP engine production had not stopped, but the rate for 2020 would fall to roughly half what it was in 2019, also indicating a rate of twenty-one airframes.

This is totally different from the five-a-month jumps Boeing and suppliers originally envisioned shortly after MAX production was cut to forty-two a month in April 2019. Some media sources were reporting that MAX-makers had eyed a rate increase to forty-seven in June, fifty-two by August and finally fifty-seven by September 2019. And of course, this was where Boeing and suppliers were poised to go to before the MAX crisis erupted.

But in February 2020 it was a 'creep, crawl, walk, jog, and then run approach,' as industry consultant Jim McAleese put it. It appeared then that the first year of 'normal' MAX production will be 2023.

In turn, Boeing and its suppliers had to adjust their earnings expectations to account for both the lack of previously planned deliveries as well as new costs, since they were positioned for fifty-seven a month in 2019. For its part, Boeing added $9.2 billion to its summary of MAX-related financial charges as part of its latest earnings report.

Cost estimates from the supply chain were by then trickling in, and they ranged from mild to eye-popping. Hayes says the production pause was projected to cost UTC just $100 million per month in sales. 'We do not anticipate any layoffs,' he adds. 'I think that would be the easiest thing to do, but quite frankly, given the scarcity of talented aerospace workers out there, we're not going to be laying anybody off for a ninety-day delay here. I think we're going to work on the backlog.'

But Spirit has already announced layoffs of at least 2,800 workers in Wichita. Increasingly, smaller suppliers are warning Wall Street of bad news, too. British aerostructures provider Senior - more formally known Senior Aerospace BWT - had said its aerospace revenue could drop twenty per cent compared with last year, due to the MAX production halt. The company, once known as Baxter Woodhouse & Taylor, became involved in the development of pressure suits and helmets for high altitude flying and with the US Space agency for space suits. In the 1950s and 1960s, the company became increasingly involved in the provision of insulation and other products for passenger aircraft. In 1958 Baxter Woodhouse & Taylor joined the Society of British Aircraft Companies and began exhibiting their products at the Farnborough Air Show. In 1999 BWT was acquired by Senior plc and become known as Senior Aerospace BWT. In April 2002 the company relocated to a 85,000 square foot purpose-built facility in Adlington, UK. Senior's operating margins were expected to slip as well, and according to media reports the company is saying the recently announced sale of its aerostructures business by Lazards might have to be shelved until more MAX certainty emerges.

Although the 737 MAX crisis is Boeing's biggest problem, other issues are dogging the company. Internal documents show it plans to record a charge for the Starliner space capsule it is developing for NASA. Starliner will need an additional uncrewed mission after it failed to reach the International Space Station as planned. The amount of the charge was not indicated. It also said that global trade tension is putting pressure on the widebody commercial jet market, Boeing's most profitable product. And it said the slowdown in production of the 787 Dreamliner widebody will stretch through at least 2023.

Will they scrap it - and what could replace it?

For as much as a good case can be made that the MAX should never have been built in the first place, I find it impossible to think that hundreds of airliners that have been built would be unceremoniously scrapped. That is just not going to happen.

So-called financial 'experts' - expert, of course, being a word from two Latin ones, 'ex' being a has-been and 'spurt' being a drip under pressure - keep running their computer models and suggest that Boeing will have to obtain eight, nine, ten or is it thirteen billion dollars worth of loans from the banks and Wall Street to tide them over and pay the veritable blizzard of compensation claims that will come. The numbers change almost daily and depend on which 'talking head' you listen to on the specialist news channels. I do not have a crystal ball, and neither do they. Still, my money is on Boeing modifying the airframe, install more sensors, change the MCAS software, issue directives for airlines to provide pilot simulator training, and then re-brand the MAX as something different. As much as the corporate strategists, public relations managers and legal eagles would like to deny all liability, I just cannot see it going to happen this time.

In the longer term, they will come up up with a completely new single-aisle design - but I do not think that will be all.

Turn back a few pages and look again at the extracts from the Fourth Quarter Reports. They show not only the dramatic downturn in business, but also just how strong Boeing is, despite all the recent problems.

As to a replacement, this is something that is notoriously difficult to predict - we're back talking about crystal balls! I still remember the many millions of words written about the famed Boeing Sonic Cruiser that never appeared!

Firstly, the Boeing 797 should be considered - it is made for what is known as the middle of the market. This is a special range of 200-250 passengers operating on dense, short-haul routes. These routes are currently flown by Boeing 737s and 757s, and airlines need a way to carry as many passengers as possible without having to deploy bigger aircraft like domestic 777s and A330s.

Thanks to the launch of the new Airbus A321XLR, airlines looking for a long-range Boeing 757 replacement are well taken care of by Airbus. In simple terms, the A321XLR can carry 220 passengers to a range of 8,700 nautical miles, while the Boeing 797-6 can carry 228 passengers to a range of 8,500 nautical miles. The Airbus A321XLR is more flexible for routes with its more extended range, and it can take around the same passengers. However, the A321XLR has one crucial advantage over the Boeing - it exists.

Thus any customers who wanted the 797 for smaller capacity long-range routes are buying the XLR, as they would get it far sooner than a 797. And they are ordering the type in bulk.

As for the bigger 797-7 with its greater passenger capacity, we only need to look at the A330-800 NEO. As in, the lack of sales for the A330-800 NEO in particular. The A330-800 NEO fills the upper mission profile of the 797-7 - although likely for a much higher price - but is not very popular at all with airlines. It is highly probable that Boeing is reluctant to offer this aircraft, especially when they could encourage airlines to buy the cheaper and slightly better Boeing 787-8 instead.

Lastly, consideration must be made regarding the background changes at Boeing. With the Boeing 737 MAX disaster having severe repercussions for the short-haul line, Boeing might scrap the 797 programme to create a new version of the 737 instead.

Boeing has already started studying the certification challenges of its Transonic Truss-Braced Wing (TTBW) ultra-efficient airliner concept as part of a new phase of work following wind tunnel tests that prove the fundamental viability of the 737-class aircraft design for typical airliner cruise

The Boeing Sonic Cruiser.

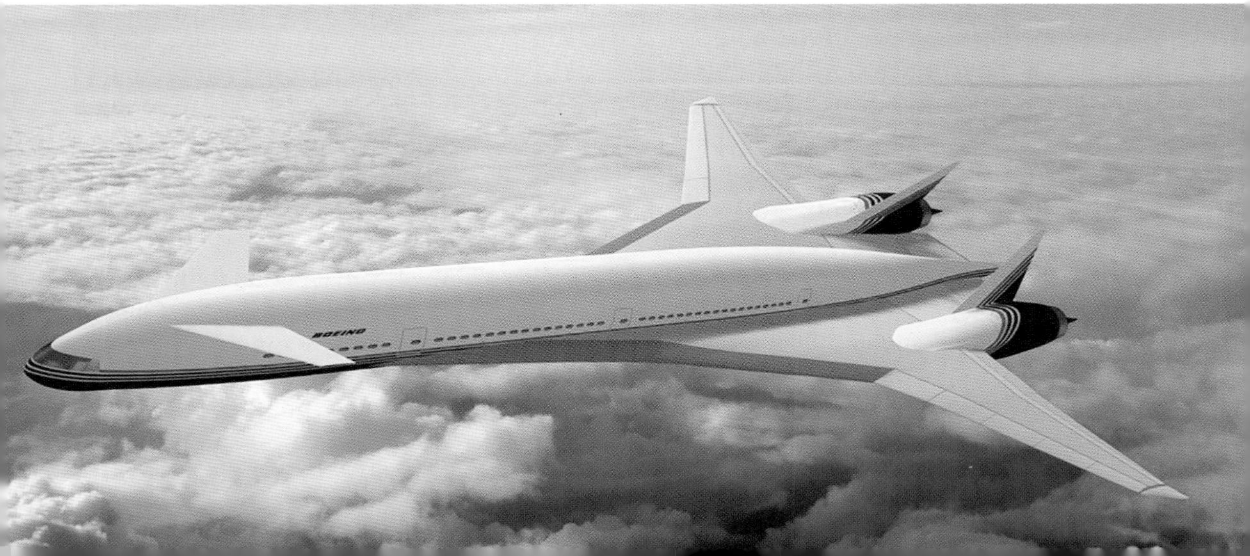

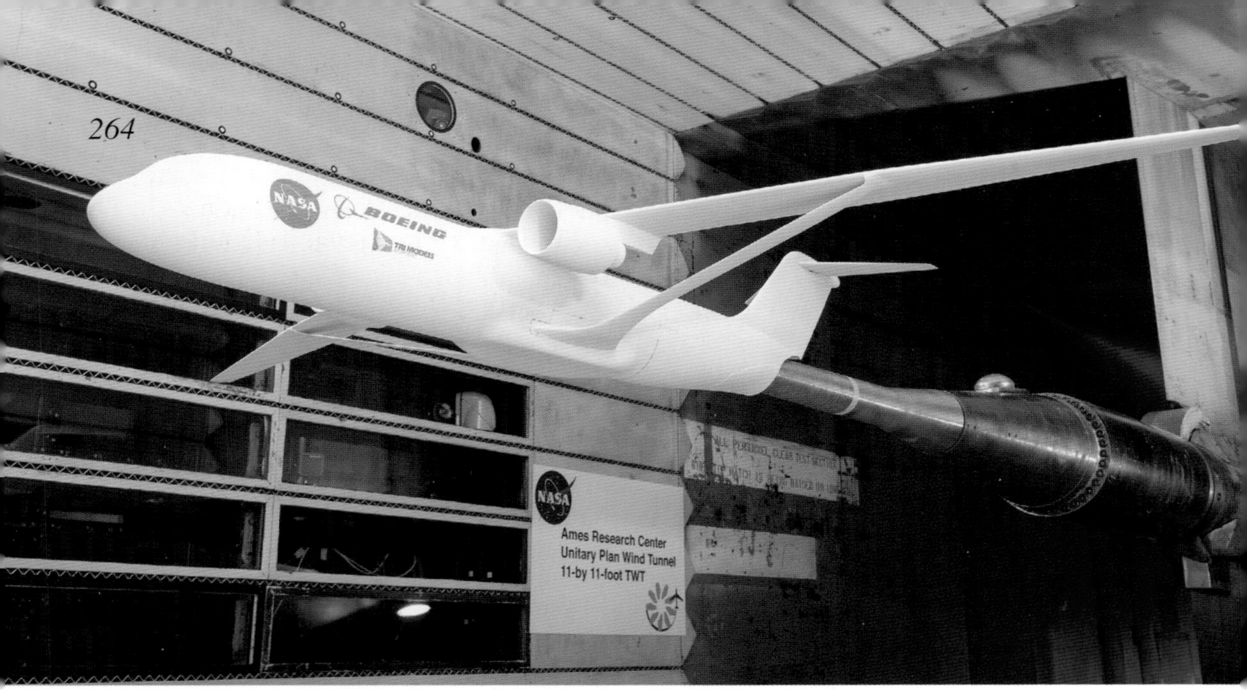

The NASA/Boeing Subsonic Ultra Green Aircraft Research project model in the Wind Tunnel at NASA Ames. *(NASA)*

speeds of Mach 0.8.

Evaluation of the slender, low-drag wing concept continues as Boeing considers technologies for a potential next-generation single-aisle design to succeed the 737 MAX towards the end of the decade. With a one hundred and seventy-foot span wing, the TTBW was originally developed in 2010 under the Boeing and NASA Subsonic Ultra Green Aircraft Research (SUGAR) programme to study new configurations for ultra-efficient airliners that could be in service from 2030 onwards.

Under the most recent phase of the programme Boeing developed and tested a high-lift system for the vehicle's long, thin wing which is braced by trusses to minimize the weight penalty of the huge span. The company estimates that thanks to the reduced induced cruise drag of the high-aspect-ratio wing, the TTBW will have a nine per cent fuel-burn advantage over a conventional tube-and-wing airliner on ranges up to 3,500 nautical miles.

The latest phase includes analysis of the aeroelastic behaviour of the configuration - which incorporated changes to the wing sweep angle as well as the position, shape and make-up of the main and jury struts. Wing sweep was increased to enable cruise speed to rise from Mach 0.745 to Mach 0.8. This caused changes to the aircraft's centre of gravity, so the wing root was moved forward, and the inboard strut moved aft and no longer positioned directly beneath the wing.

The new strut has increased chord at the fuselage, forward sweep and tapers toward the junction with the wing. Together with the 'unstacking' of the wing and strut, the changes reduced aerodynamic interference and, during wind tunnel tests, were found to have several other benefits. These include reduced compressibility drag from an improved cross-sectional area distribution, protection of the flow around the inboard strut during high-lift operations and a higher strength wing-strut structural arrangement.

By creating this increase in wing-sweep, there was a dramatic increase in vehicle wing aspect ratio, which resulted in a significant decrease in induced drag. All of this appears to have a knock-on effect with efficiency from the strut-braced configuration itself, including a significant decrease in wing bending moment which in turns leads to the potential for simplified structural attachments such as hinge joints for wing attachments.

While the design shows potential for significant performance improvements, there is no free lunch. A lot of technology has to be demonstrated and matured, including the wing-strut join. There are also other challenges such as certification for icing, ditching and crash-worthiness.

The tests also showed that the design revisions improved high-speed efficiency. High-speed testing and validation took place in NASA Ames' eleven-foot Unitary Plan Wind Tunnel starting in July 2019, while the low-speed design was tested in the NASA Langley 1 fourteen by twenty-two-foot subsonic tunnel beginning in September 2019. The transonic tests assessed vehicle lift-drag performance, longitudinal and lateral stability characteristics and included a preliminary assessment of flight controls effectiveness.

The low-speed tests, which ran through November 2019, used a fourteen-foot span model and evaluated several high-lift designs. Overall results showed beneficial changes from going to a higher sweep angle and that the design is well suited to efficient operations. Future work, which will be covered under a phase expected to be agreed with NASA in the 2020 second quarter, will focus on further maturing the concept and reducing risk. Key study areas will include investigations of high-speed buffet, alternate high-lift systems development, transonic wind tunnel tests and development of roadmaps for demonstrating structural concepts and key acoustic technologies.

Certification challenges investigated included damage tolerance to bird-strike and crash-worthiness, ditching characteristics and icing effects such as accretion, aerodynamic effects and protection.

With its Commercial Aviation division caught in the middle of the 737 MAX crisis, Boeing saw its financial results severely affected in 2019. However, the revenue of both its Defense, Space & Security and Global Services divisions are on the rise. Given the astronomic figures involved, it is not beyond the realms of possibility that the financial institutions backing Boeing could quietly insist that the company split up its activities into separate entities in order to ring-fence so as to protect them.

In its financial results for the first three quarters of 2019, Boeing reported that the revenues of the Commercial Aircraft division decreased by $16 billion compared to the same period of 2018, resulting in a operational loss of $3.8 billion against a profit of $5.2 billion a year earlier. This loss was mainly blamed on the global grounding of the 737 MAX following the two crashes.

Nevertheless, the company still managed to offset the Commercial Airplanes losses and earn $374 million net over the period. That feat was made possible by its Defense, Space & Security segment, as well as the Global Services division, which provided services to both Boeing's commercial and defence customers worldwide. The two divisions earned $2.57 billion and $2 billion respectively in the first three quarters of 2019, a 191 per cent and twelve per cent growth compared to the previous year.

COVID-19

Then the virus hit with dramatic results. On 3 April 2020 came news from Avolon, an aircraft leasing company based in Dublin, Ireland. It was founded in May 2010 by Dómhnal Slattery and a team from RBS Aviation Capital including John Higgins, Tom Ashe, Andy Cronin, Simon Hanson and Ed Riley with an initial capital of US$1.4 billion. The equity commitment of

US$1.4 billion was from four leading international investors: Cinven, CVC Capital Partners, Oak Hill Capital Partners and the Government of Singapore Investment Corporation. Between 2010 and 2014, Avolon also raised US$6.1 billion in debt from the capital markets and a range of commercial and specialist aviation banks including Wells Fargo Securities, Citi, Deutsche Bank, BNP Paribas, Credit Agricole, UBS, DVB, Nord LB and KfW IPEX-Bank. In 2017, Avolon entered the public debt markets and raised a total of over US$9 billion in debt finance. In November 2018, Avolon announced that Japanese financial institution, ORIX Corporation had acquired a 30 per cent stake in the business from its shareholder Bohai Capital, part of China's HNA Group. In April 2019, Avolon announced it had raised US$2.5 billion of additional unsecured debt which resulted in an investment-grade credit rating from Fitch, Moody's and S&P Global.

As of 31 March 2020, Avolon had an owned, managed and committed fleet of 855 aircraft but on 3 April Avolon announced that it had cancelled orders for seventy-five 737 MAXs worth $8 billion at list value in what proved to be the start of a long-expected flood of deferrals and cancellations as the coronavirus pandemic decimated air travel.

This was a clear indication of the greater risks Boeing was facing. As the grounding of the 737 MAX stretched past the first anniversary of the grounding, more and more customers were able to walk away from orders without penalty due to material adverse change clauses in sales contracts that kicked in if Boeing failed to deliver within a year of the agreed date. Also, compared to Airbus, more of Boeing's backlog was made up of orders from airlines that committed to buy more aircraft than they needed before the downturn began, analysts estimated.

The House Committee on Transportation and Infrastructure Majority Staff Preliminary Investigative Findings.

On 6 March 2020 the House Committee on Transportation and Infrastructure Majority Staff released its preliminary findings. The report outlined technical design failures on the aircraft and Boeing's lack of transparency with aviation regulators and its customers as well as Boeing's efforts to obfuscate information about the operation of the aircraft.

As Chair Peter DeFazio said. 'Our Committee's investigation will continue for the foreseeable future, as there are a number of leads we continue to chase down to better understand how the system failed so horribly. But after nearly twelve months of reviewing internal documents and conducting interviews, our Committee has been able to bring into focus the multiple factors that allowed an unairworthy airplane to be put into service, leading to the tragic and avoidable deaths of 346 people'

The Committee's preliminary findings identified five central themes that affected the design, development, and certification of the 737 MAX and FAA's oversight of Boeing. Acts, omissions, and errors occurred across multiple stages and areas of the development and certification of the 737 MAX. The Committee's investigation focused on five main areas:

1) **Production Pressures.** There was tremendous financial pressure on Boeing and the 737 MAX programme to compete with Airbus' A320neo aircraft. Among other things, this pressure resulted in extensive efforts to cut costs, maintain the 737 MAX programme schedule, and not slow down the 737 MAX production line. The Committee's investigation identified several instances where the desire to meet these goals and expectations jeopardized the safety of the flying public.

2) **Faulty Assumptions.** Boeing made fundamentally faulty assumptions about critical technologies on the 737 MAX, most notably with MCAS. Based on incorrect assumptions, Boeing permitted MCAS software designed to automatically push the plane's nose down in certain conditions; to rely on a single angle of attack (AOA)sensor for automatic activation and assumed pilots, who were unaware of the system's existence in most cases, would be able to mitigate any malfunction. Partly based on those assumptions, Boeing failed to classify MCAS as a safety-critical system, which would have offered greater scrutiny during its certification. The operation of MCAS also violated Boeing's own internal design guidelines established during development.
3) **Culture of Concealment.** In several critical instances, Boeing withheld crucial information from the FAA, its customers, and 737 MAX pilots. This included hiding the very existence of MCAS from 737 MAX pilots and failing to disclose that the AOA disagree alert was inoperable on the majority of the 737 MAX fleet, despite having been certified as a standard cockpit feature. This alert notified the crew if the aircraft's two AOA sensor readings disagreed, an event that occurred only when one was malfunctioning. Boeing also withheld knowledge that a pilot would need to diagnose and respond to a 'stabilizer runaway' condition caused by an erroneous MCAS activation in 10 seconds or less, or risk catastrophic consequences.
4) **Conflicted Representation.** The Committee has found that the FAA's current oversight structure with respect to Boeing created inherent conflicts of interest that jeopardised the safety of the flying public. The Committee's investigation documented several instances where Boeing authorized representatives failed to take appropriate actions to represent the interests of the FAA and to protect the flying public.
5) **Boeing's Influence Over the FAA's Oversight.** Multiple career FAA officials have documented examples to the Committee where FAA management overruled the determination of the FAA's own technical experts at the behest of Boeing. In these cases, FAA technical and safety experts determined that certain Boeing design approaches on its transport category aircraft were potentially unsafe and failed to comply with FAA regulation, only to have FAA management overrule them and side with Boeing instead. These five recurring themes paint a disturbing picture of Boeing's development and production of the 737 MAX and the FAA's ability to provide appropriate oversight of Boeing's 737 MAX program. These issues must be addressed by both Boeing and the FAA in order to correct poor certification practices that have emerged, faulty analytical assumptions that have surfaced, notably insufficient transparency by Boeing, and inadequate oversight of Boeing by the FAA.

The Committee's preliminary investigative findings that were grouped into six distinct categories:

1) **FAA Oversight –** The FAA failed in its oversight responsibilities to ensure the safety of the travelling public.
2) **Boeing Production Pressure –** Costs, schedule, and production pressures at Boeing undermined safety of the 737 MAX.
3) **MCAS –** Boeing failed to appropriately classify MCAS as a safety-critical

system, concealed critical information about MCAS from pilots, and sought to diminish the focus on MCAS as a 'new system' in order to avoid greater FAA scrutiny and increased pilot training requirements.

4) **AOA Disagree Alert –** Boeing intentionally concealed information from the FAA, its customers and pilots about inoperable AOA Disagree alerts installed on most of the 737 MAX fleet, despite their functioning being 'mandatory' on all 737 MAX aircraft, and the FAA has failed to hold Boeingaccountable for these actions.
5) **737 MAX Pilot Training –** Boeing's economic incentives affected the company's transparency with the FAA, customers, and 737 MAX pilots regarding pilot training requirements.
6) **Response to Lion Air Accident -** Both Boeing and the FAA gambled with the public's safety in the aftermath of the Lion Air crash, resulting in the death of 157 more individuals on Ethiopian Airlines flight 302 less than five months later.

Conclusion: These preliminary investigative findings make clear that Boeing must create and maintain an effective and vigorous safety culture and the FAA must develop a more aggressive certification and oversight structure to ensure safe aircraft designs and to regain the confidence of the flying public. We hope these preliminary findings will help pave the way for legislative reforms as the Committee's investigation continues to identify the actions and events that undermined the design, development, and certification of the 737 MAX aircraft and led to the tragic death of 346 people.

Is a bail out on the cards?

By mid-March 2020 reports started to circulate in the media that Boeing was planning on fully utilising its recently inked $13.8B loan facility. On 18 March the company confirmed that it had done just that.

That same day reports also circulated Boeing has told the government it needed significant support to meet its liquidity needs and was seeking billions in loan guarantees and other assistance, for the company's stock plummeted nearly 18 per cent the previous day amid growing worries about how the aircraft maker and defence contractor would weather a series of recent storms.

Clearly, Boeing thought of this as more an industry bailout, noting that seventy per cent of its revenue typically flows through to suppliers. Boeing asked the government should take some form of an ownership position in the company in exchange for a big slug of taxpayer's cash.

Boeing stressed that the long-term outlook for the industry remained strong, but stressed that it needs immediate relief until the coronavirus crisis passes. Boeing's shares shed over sixty per cent of its value in the previous month.

President Trump indicated that Boeing would get financial assistance and said that he considered it the best company in the world before it came under scrutiny over the safety of its 737 MAX aircraft. 'Yes, I think we have to protect Boeing. We have to absolutely help Boeing,' Trump said at the White House. 'Obviously when the airlines aren't doing well then Boeing is not going to be doing well. So we'll be helping Boeing'. This then begs the question - what would the Trump Administration do?

Overall the US government - including even the Pentagon - cannot be happy with Boeing's performance, not withstanding the current COVID-19 crisis. Boeing's shares were dropping like a rock well before the coronavirus mess. Its management was in disarray and had lost the confidence of its workers, its clients and the public. The 737 MAX disaster still is unfolding as I write this and is not yet settled. Few people in the industry have any real confidence that all the problems of the MAX have been isolated or addressed, and some airlines are stuck with airliners they can't use and which instils fear and dread in their customers.

The United States Coronavirus Aid, Relief, and Economic Security Act (the 'CARES Act') package Congress passed on 27 March 2020 included $454 billion in loans for businesses, including $17 billion for businesses critical to maintaining national security, like Boeing.

However, on 8 April, it became clear that CEO David Calhoun was considering forgoing the relief Boeing requested. It seemed that Calhoun was leary to accept the aid if that meant the government could take an ownership stake in the company.

Boeing had plans in place to pay employees through Tuesday and then 30,000 employees will have to use paid time off. Boeing suspended operations at two Washington state plants - Puget Sound and Moses Lake - until further notice. It also temporarily suspended operations in South Carolina and operations at its facilities in Ridley Township, Pa.

The case for Boeing is its economic and military importance. As the song *Love and Marriage* put it, '*you can't have one without the other*'. Or can you?

One option the government could consider is requiring that the military business be separated, stock listed separately, or completely privatised. One presumes that the military part of the company is still doing fairly well. Moreover, the USAF is still accepting defective air refueling tankers from Boeing, even though they are full of FOD buried inside the air frames by sloppy workers and a key component, the refuelling boom, does not work properly. While not explained, it is likely that the Air Force desperately needs the tankers and also decided to throw a bone to Boeing and allow them to fix their mistakes as a gesture to the hard times the company is enduring.

The military side of Boeing could be put up for sale. While the company depends on 'green aircraft' from Boeing (being the only commercial aircraft company that has survived the brutal competition in the United States, a competition that saw the end of both Lockheed's commercial jet business and the acquisition by Boeing of McDonnell Douglas), an independent military arm could buy aircraft from anywhere and on whatever terms the government deems acceptable. Separation would safeguard the national security interest of the American people.

On the commercial side, the US government should demand oversight of Boeing's operations, from management on down to factory floor workers. That oversight should be tough, unrelenting and demanding. Its members should be critics of the company, not go-along supporters and biased friends. $60 billion is a big ticket, and the taxpayer should not have to face more bailouts down the line, or shoddy workmanship, or bad management.

Certainly, the recovery of the commercial division could take a while. The future of the Boeing 737 MAX family is still very uncertain, and some of its other coming programmes are struggling. The 777X, yet to enter service, was recently affected by several development setbacks, and customers are starting to lose interest. On 20 November 2019, Emirates officially announced it would cut a fifth of its 777X order - twenty-four aircraft - to replace them by thirty 787-9 Dreamliners, itself a product that has been hit with problems. Meanwhile, the 'New Midsize Airplane' programme has yet to be unveiled.

While they allow the manufacturer to limit the damage caused by the Boeing 737 MAX, a so-called 'spin-off scenario' in corporate-speak to preserve the financial future of the defence and services divisions should not be ruled out. Boeing Defense, Space and Security division benefits from a privileged relationship with the American military and its allies. Its last major contract dates back to September 2019, when the Pentagon ordered fifteen KC-46 Pegasus in a contract valued at $2.6 billion despite developmental problems affecting the refuelling tankers.

An independent Boeing defence company would be placed in a good position among the biggest defence contractors in the world. To compare with Boeing's $2.57 billion in net defence earnings during the first three quarters of 2019, the market leader Lockheed Martin reported $3.8 billion net over the same period.

As for Boeing Global Services, it racked up $225 billion in contracts to be spread over the next ten years at the last Dubai Airshow, in November 2019.

However, this spin-off scenario would mean gargantuan restructuring costs for a situation that might be temporary and the prospect of seeing the positions of some of the 64,000 employees threatened would almost certainly create a political scandal.

I will conclude this on a historical reminder that I mentioned in the Introduction: Boeing is already the result of an historical three-way split. In 1934, the US Congress voted the Air Mail Act which prohibited airlines and manufacturers to be under the same umbrella. The United Aircraft and Transport Corporation was thus spun off into Boeing, United Airlines and what is known today as United Technologies. So never say never!

And finally...

With every book I write, no matter how hard I try to keep things under wraps, I always get asked what I am currently working on. I was interrogated in even greater detail when it became common knowledge amongst family and friends that I was working on this one. There was but one question: did I think it was safe to fly on a 737?

The answer is simple and, believe it or not, the answer is yes. Given all that I have written here, that statement probably needs further explanation.

At the time of writing the global 737 MAX fleet is grounded, currently undergoing modification, upgrades and checks to ensure they are fit for service. Passengers are not allowed to fly in them, so there is no likelihood of travelling on one just yet.

Boeing, the US FAA and other global aviation safety authorities cannot afford any more scandals, and so when the 737 MAX is put back into service, the travelling public can rest assured that everything possible will have been investigated, checked, studied and reported on and then fixed to make sure that the aircraft is safe and that the crews have been fully trained to operate them.

As for the older versions of the 737 currently flying, although there undoubtedly have been other problems, these will also have been investigated and checked on to ensure that they are all as safe as they can possibly be.

If there is a lesson to be learned from this debacle I have unearthed, it is this. Engineers, designers, inspectors and flight crews should be allowed to do what they are trained to do and to make use of their hard-won experience. Listen to what they say and let them do their jobs! Keep the devil's triumvirate of Accountants, Public Relations Managers and 'Corporate Strategists' with their shiny, but totally useless MBAs at arms, or even flagpoles length away from the critical area of safety.

Appendix 1

Boeing Custoimer Codes

01	Piedmont
02	Northern Consolidated / Wien Air Alaska
03	Caribair
04	Britannia Airways
05	Braathens SAFE
06	K.L.M - Royal Dutch Airlines
07	Luftwaffe / West German Air Force
08	Icelandair
09	China Airlines
10	Wien Colsolidated / Wien Air Alaska
11	Wardair
12	M.S.A / Singapore Airlines
13	Ariana - Afghan Airlines
14	Pacific Southwest Airlines (PSA)
15	Lake Central
16	LAN Chile
17	CP Air / Canadian International
18	B.E.A
19	NZ National Airways Corp / Air New Zealand
20	Boeing
21	Pan American World Airways
22	United Air Lines
23	American Airlines
24	Continental Airlines
25	Eastern Airlines
26	USAF Military Airlift Transport Service (MATS)
27	Braniff Airways
28	Air France
29	SABENA
30	Lufthansa
31	T.W.A
32	Delta Air Lines
33	Air Canada
34	Transair Sweden
35	National Airlines
36	B.O.A.C / British Airways
37	Air India
38	Qantas
39	Cubana
40	Pakistan International Airlines
41	Varig
42	Nordair
43	Alitalia
44	South African Airways
45	Seaboard World Airlines
46	Japan Air Lines
47	Western Air Lines
48	Aer Lingus
49	Flying Tiger Line
50	Trek Airways
51	Northwest Airlines
52	Aeronaves de Mexico / Aeromexico
53	United States Air Force
54	Mohawk Airlines
55	Executive Jet Aviation

737-2T5 G-BHVI of Orion Airways. This aircraft's 'claim to fame' was that it was the 737th 737 built.

HS-TBB, a 737-2P5 of Thai Airways.

56	Iberia
57	Swissair
58	El Al
59	Avianca
60	Ethopian Airlines
61	F.A.A
62	Pacific Northern
63	Ghana Airways (ntu) / Faucett
64	Mexicana
65	British Eagle
66	United Arab Airlines / Egyptair
67	Cathay Pacific Airways
68	Saudia / Saudi Arabian Airlines
69	Kuwait Airways
70	Iraqi Airways
71	Trans International / Transamerica
72	Airlift International
73	World Airways
74	Iraqi Airways (ntu) / Libyan Arba (ntu)
75	Pacific Western Airlines
76	Trans Australian
77	Ansett Airlines
78	B.W.I.A
79	Saturn Airlines
80	Bankers Trust (TWA)
81	All Nippon Airways
82	T.A.P - Air Portugal
83	S.A.S
84	Olympic Airways
85	American Flyers
86	Iran Air
87	Aerolineas Argentinas
88	M.E.A. / Aerotechnik Deutsche Charter
89	Toa Domestic / Japan Air System
90	Alaska Airlines
91	Frontier Airlines
92	Air Asia / Air America
93	Pacific Air Lines / Air California
94	Syrian Arab Airlines / Syrianair
95	Northeast Airlines
96	Quebecair
97	Aloha Airlines
98	Air Zaire
99	British Caledonian
1A	Martinair Holland
1B	China Southern Airlines
1C	Government of Romania
1H	Emirates
1K	Airtours International Airways Ltd.
1L	China Xinjiang Airlines
1M	Oman Air
1Q	Tombo Aviation Inc.
1R	Virgin Atlantic Airways
1S	Deutsche BA
1U	BBJ Customer
2A	Hawaiian Airlines
2C	Air UK Leisure
2J	Arab Leasing
2K	Turkmenistan Airlines / Akhal
2L	Azerbaijan Airlines
2P	Government of United Arab emirates
2Q	Ukraine International Airlines
2R	Pegasus Airlines
2T	Tracinda Corp. (BBJ Customer)
2U	JABJ / Picton II Ltd. (BBJ Customer)
3A	Ansett Worldwide (AWWAS) Nordstress Ltd.
3B	Air Mauritius
3C	Euralair International
3N	American Trans Air
3P	Uzbekistan Airways
3Q	Boeing (BBJ Customer)
3R	Western Pacific Airlines
3S	Pembroke Capital
3T	Real Estate Exchange Inc. (BBJ Customer)

3U Chartwell Aircraft Company (BBJ Customer)
3V Easyjet Airline Co Ltd.
4A United Parcel Service - UPS
4K Air Nippon
4N China United Airlines
4P Hainan Airlines
4Q Mid East Jet
4S GB Airways
4T GECAS (BBJ Customer)
4U Air Shamrock (BBJ Customer)
4V Superior International Aviation Services (BBJ Customer)
5A Presidential Airways
5B Germania
5C Xiamen Airlines
5D LOT - Polish Airlines
5E EVA Air
5F GATX Capital Corp
5H Itochu AirLease Corp.
5N Shandong Airlines
5P Air Europa
5R Jet Airways (India)
5S CSA - Czech Airlines
5T First Union Commercial Corp. (BBJ Customer)
5U Dobro Ltd. (BBJ Customer)
5V General Electric Corporation (BBJ)
6B Cal Air / Novair International Airways
6D Shanghai Airlines
6E VIVA Air
6J Air Berlin
6K Vietnam Airlines Corporation
6M Virgin Express
6N GE Capital Aviation (GECAS)
6Q Boullion Aircraft Holding
6R Wuhan Airlines
7A Far Eastern Air Transport
7B Air Holland
7C Japanesse Self Defence Force
7D Air Seychelles
7E Aeromaritime International
7G Malev - Hungarian Airlines
7K Zhongyuan Airlines
7L Shenzen Airlines
7Q Novel Leasing Company
7U Atlas Air
8A Air 2000 Limited
8B Istanbul Airlines
8E Asiana Airlines
8J Tarom
8N Chilean Air Force
8S China Xinjiang Airlines
9A ARAVCO
9B Government of Brunei
9D Linjeflyg
9H Leisure International
9J Shorouk Air
9K Shenzen Airlines
9L Air China
9M Air Austral
9N Mid East Jet
9P China Eastern Airlines
9R Pro Air
9T Usal Ltd. (BBJ Customer)
9U Wilmot Trust company (BBJ Customer)
A0 Lloyd Aereo Boliiviano
A1 VASP
A2 Modern Air
A3 Pluna
A4 Air California
A6 Essex International / LTV Capital
A7 Trans Caribbean
A8 Indian Airlines
A9 Transair Canada
B1 DETA Mozambique / LAM

373-2T4ADV EC-DUL of the Spanish charter airline Spantax.

T-43s were not common visitors to Europe. Here 31154 awaits the crowds at the start of a Mildenhall Air Show in the 1980s. *(John Hamlin)*

B2	Air Madagascar
B3	U.T.A
B4	M.E.A
B5	Korean Air
B6	Royal Air Maroc
B7	Allegheny / USAir / US Airways
B8	Austrian Airlines
C0	GATX / Boothe
C3	Cruzeiro do Sul
C9	Luxair
D1	Universal Airlines
D3	ALIA - Royal Jordanian Airlines
D4	Ozark Airlines
D6	Air Algerie
D7	Thai Airways International
E0	Dubai Air Wing (BBJ Customer)
E1	Eastern Provincial
E3	LADECO
E7	Arkia
F1	Air New Zealand (ntu)
F2	THY - Turkish Airlines
F5	Government of Portugal / Portuguese Air Force
F6	Philippine Airlines
F8	Royal Nepal Airways
F9	Nigeria Airways
G1	Government of Saudi Arabia
G4	United States Air Force
G5	L.T.S / L.T.U.
G7	America West Airlines
H2	I.T.T
H3	Tunis Air
H4	Southwest Airlines
H5	Mey Air
H6	Malaysian Airlines
H7	Cameroon Airlines
H9	J.A.T - Jugoslav Airlines
J0	Air Jamaica
J1	Dominicana
J4	Sterling Airways
J6	C.A.A.C / Air China
J7	National Aircraft Leasing
J8	Sudan Airways
J9	Iranian Air Force
K1	Tarom / Romanian Government
K2	Transavia / KLM
K3	Aviogenex
K5	Hapag Lloyd
K6	SAHSA
K9	Bavaria Flug
L4	American Capital Aviation
L5	Libyan Arab Airlines
L6	Aviation Services & Support
L7	Air Nauru
L8	Government of Yugoslavia
L9	Maersk Air
M0	ARIA - Aeroflot Russian International Airlines
M1	Pelita Air Service
M2	TAAG - Angola Airlines
M6	Royal Brunei
M7	Hughes Air West
M8	Trans European Airways
M9	Zambia Airways

N0	Air Zimbabwe
N1	Government of Venezuela
N3	Government of Brazil
N6	Government of Nigeria
N7	Government of Egypt
N8	Yemen Airways
N9	Government of Niger / Air Niger
P1	Goverment of Qatar
P3	HRH Talal bin Abdul Aziz (BBJ Customer)
P5	Thai Airways
P6	Gulf Air
P8	Bahrain Amiri Flight
Q2	Air Gabon
Q3	Southwest Air Lines / Japan TransOcean
Q4	Transbrasil
Q5	Government of Liberia / Air Liberia
Q6	LACSA
Q8	I.L.F.C
Q9	Itel Corp.
R1	Government of Cameroon
R4	Alyemda
R6	Air Guinee
R7	Cargolux Airline International
R8	Air Tanzania
S1	TACA International Airlines
S2	Federal Express
S3	Air Europe
S4	Air Afrique
S5	Eldorado Aviation
S7	North Central Airlines / Republic Airlines
S9	Maritime Investment
T0	Texas Air Corp / Continental Airlines
T2	Dome Petroleum
T3	Evergreen
T4	Air Florida
T5	Orion Airways
T7	Monarch Airlines
T8	Polaris Leasing
T9	Boeing
U3	Garuda Indonesia
U4	OSL Villa Holidays
U5	Government of Jordan
U8	Kenya Airways
U9	Polynesian Airlines
V2	TAME
V3	COPA Airlines
V5	Bahamasair
V6	Petrolair Systems SA
V8	Air Executive / Busy Bee
W0	China Yunnan Airlines
W2	Aerotour
W6	Government of Morocco
W8	NOGA Import
X2	Air Pacific
X3	Air Charter International
X4	Supair
X6	Markair
X8	Wistair Corp
X9	Indonesian Air Force
Y0	G.P.A Group
Y4	Rafic B Hariri (Saudi customer)
Y5	Air Malta
Y9	Air Malawi
Z0	China Southwest
Z5	United Arab Emirates Royal Flight
Z6	Royal Thai Air Force
Z8	South Korean Air Force
Z9	Lauda Air
AB	AB Airlines
AD	Eastwind Airlines
AF	United States Navy / Marines
AH	General Electric Capital Corp. (BBJ Customer)
AJ	BBJ Customer
AK	Privitair (BBJ Customer)
AL	Singapore Aircraft Leasing Enterprise (SALE)
AN	Saudi OGER (BBJ Customer)
AR	Taiwan Air Force
AS	Ryanair
AV	Newsflight II Inc. (BBJ Customer)
AW	A.S. Bugsham & Bros. (BBJ Customer)
AX	ARAMCO
BC	Boeing NetJets (BBJ Customer)
BD	AirTran Airways Inc.
BF	Funair Corp. (BBJ Customer)
BG	Flightlease AG
BH	BBJ Customer
BJ	Atlas Air (BBJ Customer)
BK	CIT Leasing Corp.
BL	Midwest Airlines
BQ	GECAS
BS	North American Airlines
BX	Midway Airlines
CB	Hawaiian Airlines
CG	GKW Aviation LLC (BBJ Customer)
CJ	BBJ One Inc. (BBJ Customer)
CM	Aerolineas de Baleares Aebal
CN	Matela Offshore Ltd. (BBJ Customer)
CP	Ford Motor Company (BBJ Customer)
CQ	JMC Airlines Limited
CT	Westjet
CU	Tutor-Saliba corp. (BBJ Customer)
CX	GATX - Flightlease
DF	Royal Australian Air Force (BBJ Customer)
DM	United States Air Force (BBJ Customer)

Captain Roger Cooper brings in British Airways' 737-436 G-DOCV *'River Thurso'* complete with the Benyhone tartan on the fin into land at London Gatwick. *(author)*

DO	BBJ Customer
DP	BBJ Customer
DR	Multiflight (BBJ2 Customer)
DT	Royal Australian Air Force (BBJ Customer)
DV	Lowa Ltd. (BBJ2 Customer)
DW	Bausch & Lomb, Inc. (BBJ Customer)
DX	Kazakhstan Airlines
EA	Azteca Airlines
EC	Dubai Air Wing (BBJ2 Customer)
ED	BBJ Customer
EE	Air Senegal
EF	BBJ2 Customer
EG	Samsung Aerospace (BBJ Customer)
EH	GOL Airlines
EI	BBJ Customer
EJ	Grupo Omnilife S.A. de C.V. (BBJ Customer)
EL	Swiflite Aircraft Corporation (BBJ Customer)
EM	BBJ Customer
EQ	BBJ2 Customer
ES	Royal Australian Air Force/Turkish Air Force (BBJ Customer)
ET	BBJ Customer
EV	Jade Cargo International
EX	BBJ2 Customer
EY	Italian Air Force (B767 Tanker)
EZ	United States Air Force (B767 Tanker)
FB	BBJ Customer
FD	BBJ Customer
FE	Virgin Blue Airlines Pty Ltd.
FG	Saudi Ministry Of Finance & Economy (BBJ Customer)
FH	RBS Aviation Capital
FK	Japan Air Force (B767 Tanker)
FT	Air China Cargo
FX	Etihad Airways
GC	BBJ Customer
GG	BBJ2 Customer
GJ	Spicejet Ltd.
GK	Buraq Air Transport
GP	Lion Air
HA	TNT Airways, S.A.
HG	Air India Express
HI	Indian Air Force (BBJ Customer)

Bibliography

If I listed every reference to every small item I used in this work from assorted newspapers and aviation magazines, the bibliography would be more significant than the book!

Therefore the reader will have to be content with my listing of those publications and the years that have been used and then if needed, search for themselves!

Firstly, the newpaper archives of

The Daily Express (UK) 2001 - 2020
The Daily Mail (UK) 2001-2020
The Guardian (UK) 2001-2020.
The Hill (USA) 2001-2020.
The Independant (UK) *2001-2020.*
The New York Times (USA) 2001-2020.
The Seattle Times (USA) 2001-2020.
The Times (UK) 2001-2020.
The Washington Post (USA) 2001-2020.

A similar occurance happens with the national and international aviation magazines, especially where they have ran tiny news features.

Air Pictorial: 1965-2002. a British aviation magazine covering contemporary and historical military and civil aviation topics.

Aircraft Illustrated: 1968-2009. A British monthly aviation magazine covering military and civil aviation topics with an emphasis on photographic features. It was first published in 1968.

Airliner World: 1999 to date. An aviation magazine published by Key Publishing in Stamford, Lincolnshire, United Kingdom and distributed by Seymour Distribution Ltd in London. In the United States, the magazine is distributed from Key Publishing's office in Avenel, New Jersey.

Airliners: The Worlds Airline Magazine. 1988-2010. An American magazine dedicated to the airline industry. Six issues were circulated each year. The title was first published by World Transport Press in 1988. Since the 100th issue (July/August 2006) it was produced by Airliners Publications, LLC.

Airlines: IATA's multi-platform magazine, offering the latest aviation business news, exclusive airline CEO interviews, and expert insight and analysis.

Airways Magazine: commercial aviation source of news, photos, knowledge, and special events.

Aviation News: 1972-2002. The magazine was founded in 1972 as a fortnightly tabloid newspaper-style publication by Alan W. Hall (Publications) Ltd. In July 1983 (Volume 12 Number 1) it changed to an A4 format magazine.

Aviation Week & Space Technology: The Aviation Week Network also publishes *Business & Commercial Aviation* and *Air Transport World.*

Flight International: 1965 to date.

Boeing Magazine: 1964 (12 editions)
Boeing Magazine: 1965 (12 editions)
Boeing Magazine: 1966 (12 editions)
Boeing Magazine: 1967 (12 editions)
Boeing Magazine: 1968 (12 editions)
Boeing Magazine: 1969 (12 editions)
Boeing Magazine: 1970 (12 editions)
Boeing Magazine: 1971 (12 editions)
Boeing Magazine: 1972 (12 editions)
Boeing Magazine: 1973 (12 editions)
Boeing Magazine: 1974 (12 editions)
Boeing Magazine: 1975 (12 editions)

The archives of the National Transportation and Safety Board (USA).

The archives of the Federal Aviation Authority(USA).

The House Committee on Transportation and Infrastructure has jurisdiction over all aspects of civil aviation, including safety, infrastructure, labor, and international issues. Following two accidents overseas involving Boeing 737 MAX aircraft that killed 346 people and led to the worldwide grounding of the aircraft, the Committee launched an investigation to ensure accountability, transparency in the certification process, and most importantly the safety of the traveling public.

Releases to the media (In chronological order)

- Chair DeFazio Statement on Crash of Ethiopian Airlines Flight 302
- Chairs DeFazio, Larsen Respond to Grounding of Boeing Aircraft
- Chairs DeFazio, Larsen Request DOT IG Investigation into FAA Certifying Boeing 737 MAX
- Chair DeFazio Encourages FAA/Boeing Employees to Utilize Whistleblower Page
- Chairs DeFazio, Larsen Urge FAA to Conduct Third-Party Review into Boeing 737-MAX
- As Part of Investigation into Boeing 737 MAX Certification Process, Committee Sends Records Requests to FAA, Boeing
- Chairs DeFazio and Larsen Applaud FAA

Announcement Ordering a Third-Party Review into Boeing 737 MAX
- Chairs DeFazio, Larsen Statements from Hearing on 'Status of the Boeing 737 MAX'
- Chairs DeFazio and Larsen Seek Answers About Delayed Notice of Defective Angle-of-Attack Alert on 737 MAX
- Chairs DeFazio, Larsen Announce Aviation Subcommittee Hearing on 'Status of the Boeing 737 MAX: Stakeholder Perspectives'
- Chairs DeFazio, Larsen Statements from Hearing on 'Status of the Boeing 737 MAX: Stakeholder Perspectives'
- Chairs DeFazio, Larsen Announce Aviation Subcommittee Hearing on Safety
- Chairs DeFazio, Larsen Statements from Hearing on 'State of Aviation Safety'
- As Part of the Committee's Investigation into the Boeing 737 MAX, Chairs DeFazio and Larsen Formally Request Interviews with Several Boeing Employees
- Chair DeFazio Formally Invites Boeing CEO Dennis Muilenburg to Testify at Hearing in October on the 737 MAX
- Chair DeFazio Announces Boeing CEO Dennis Muilenburg Will Testify at Hearing in October on the 737 MAX
- Chair DeFazio, Larsen React to NTSB Aviation Safety Recommendations
- Chairs DeFazio, Larsen Statement on Newly-Released Review of the Certification of the Boeing 737 MAX
- Chair DeFazio Blasts Boeing about Newly-Revealed Messages: This is not about one employee, this is about a failure of safety culture
- In New Letter to Transportation Secretary Chao, Chair DeFazio Sharply Questions Why Outrageous Emails Related to the Boeing 737 MAX Are Only Now Being Revealed, Months after Committee's Initial Request
- Chairs DeFazio, Larsen Statement on Lion Air Crash Report
- Chairs DeFazio, Larsen Statements from Hearing titled, "The Boeing 737 MAX: Examining the Design, Development, and Marketing of the Aircraft"
- Amid Committee's Ongoing Investigation into the Certification of the 737 MAX, Chairs DeFazio and Larsen Raise New and Serious Concerns to FAA About Other Safety-Related Issues
- Following Recent Hearing on the Boeing 737 MAX, Chair DeFazio Presses Boeing CEO for Additional Information About Decisions on MCAS, Grounding the Aircraft, CEO Pay, Boeing's Legal Strategy and More
- Chair DeFazio Announces Hearing on FAA Oversight of the Boeing 737 MAX Certification
- Chairs DeFazio, Larsen Statements from Hearing titled, "The Boeing 737 MAX: Examining the Federal Aviation Administration's Oversight of the Aircraft's Certification"
- Chair DeFazio statement on Boeing's Announcement Regarding Simulator Training for 737 MAX pilots
- Chair DeFazio's Statement on Newly-Released Boeing Emails
- Chairs DeFazio, Larsen Release Statements on Newly-Released Report on Certification Process for the Boeing 737 MAX
- Nearly One Year After Launching Its Boeing 737 MAX Investigation, House Transportation Committee Issues Preliminary Investigative Findings Letters
- T&I Leaders Call for Investigation of Training Provided to Foreign Pilots
- Chairs DeFazio, Larsen Request DOT IG Investigation into FAA Certifying Boeing 737 MAX
- Chairs DeFazio, Larsen Press DOT, FAA on Records Request Delay
- Chairs DeFazio, Larsen Send Letter to Colleagues about the Committee's Boeing 737 MAX Investigation
- Following Recent Hearing on the Boeing 737 MAX, Chair DeFazio Presses Boeing CEO for Additional Information About Decisions on MCAS, Grounding the Aircraft, CEO Pay, Boeing's Legal Strategy and More

Hearings
May 15, 2019: Aviation Subcommittee Hearing: 'Status of the Boeing 737 MAX'

June 19, 2019: Aviation Subcommittee Hearing: 'Status of the Boeing 737 MAX: Stakeholder Perspectives'

July 17, 2019: Aviation Subcommittee Hearing: 'State of Aviation Safety'

October 30, 2019: Full Committee Hearing: 'The Boeing 737 MAX: Examining the Design, Development, and Marketing of the Aircraft'

December 11, 2019: Full Committee Hearing: 'The Boeing 737 MAX: Examining the Federal Aviation Administration's Oversight of the Aircraft's Certification'

Manuals.
Boeing 737-200 Maintenance Training Manual.
Boeing 737-300 Maintenance Training Manual.
Boeing 737-400 Maintenance Training Manual.
Boeing 737 MAX-8 Maintenance Training Manual.

Index

Lufthansa's 737-130 D-ABEB *City Jet 737 Regensburg* is about to complete another service into London Heathrow *(author)*

N501NG was a 737-2T5, originally delivered to Orion Airways. It is seen here while being leased from International Leasing and Finance Corp by NICA. *(Hank Warton Collection)*

C

JA8453 of All Nippon Airways was a 737-281. *(Bob LeHat Collection)*

5B-DBY, a 737-31S of Helios Airways named *Olympia*. When operating as Helios Flight 522, on 14 August 2005, the crew became incapacitated due to the incorrect setting of the air craft's pressurisation system on a flight from Larnaca to Prague, in the Czech Republic with an intermediate stop at Athens, Greece. As the airliner climbed over the Mediterranean, the cabin altitude alert horn sounded. The crew possibly thought it was an erroneous takeoff configuration warning because the sound is identical. The aircraft eventually crashed into hills near Grammatiko, 40 km from Athens, killing all 121 passengers and crew on board.

Kulula is a South African low-cost airline, operating on major domestic routes from O. R. Tambo International Airport and Lanseria International Airport, both serving the city of Johannesburg.The name 'Kulula' comes from the Nguni languages of Zulu and Xhosa, meaning It's easy

OK-PIK - painted as 'flying 101'in this set - is a 737-86N, and clearly demonstrates that the airline very much has a sense of humour, covering the entire fuselage with somewhat humorous explanations as to what goes where!

ZS-OAP (above) is a 737-4S3 and is painted as 'flying 102'.
(all photos Kulula.com)

Q

R

S

T

U

V

W

Y

Z